va

John G. Knighton.

Rare

PUBLIC LIBRARY OF CINCINNATI AND HAMILTON COUNTY

Authorized in 1853 as a Public and School Library, it became in 1867 an independent Public Library for the city, and in 1898 the Library for all of Hamilton County. Departments in the central building, branch libraries, stations and bookmobiles provide a lending and reference service of literary, educational and recreational material for all.

The Public Library is Yours . . . Use it!

THE

PYROTECHNIST'S COMPANION.

THE

PYROTECHNIST'S COMPANION:

OR,

A FAMILIAR SYSTEM

OF

RECREATIVE FIREWORKS.

BY

G. W. MORTIMER.

First American,

FROM THE SECOND LONDON REVISED EDITION.

PHILADELPHIA:

HENRY CAREY BAIRD.

1852.

PHILADELPHIA:
T. K. AND P. G. COLLINS, PRINTERS.

PREFACE.

The Introduction prefixed to the following little work, supersedes the necessity of an extended Preface, and leaves little more to be mentioned than its design and occasion.

The design of it is to be a useful assistant to those who are fond of a rational and scientific amusement, and the occasion of it arises from the great scarcity of, and general difficulty of procuring, any work on the subject; none having appeared, worthy of notice, since that published by Lieutenant Robert Jones, in 1760, and those by the French artists mentioned in the Introduction.

In didactic particulars, the author has occasionally availed himself of the language of the best writers, where such has been corroborated by subsequent experience.

Perspicuity has been a particular object through the work, and when technical terms have been used they are generally followed by familiar explanations, and the author feels assured that the whole will be found perfectly intelligible to every reader. To experienced pyrotechnists, this little work cannot be expected to afford much additional information, yet it may contain some few particulars not known to them before, which, from their practical utility, it is hoped will prove acceptable.

The author trusts that this work may prove a useful assistant to those who are unacquainted with the principles of the art on which it treats: and if in any way it should contribute to this purpose, an apology for obtruding it upon the public will certainly be unnecessary.

INTRODUCTION.

The term *Pyrotechny* is derived from *pyr* and *techny*, the two Greek words for FIRE and ART; and it is the art of employing fire for purposes of utility or pleasure. The term has been applied by some writers to the use and structure of firearms, and artillery employed in the art of warfare; but in the present publication, a different view of the subject will be taken; for there cannot be any amusement in the motion of a bullet, which decimates so many human beings, nor in the action of a bomb-shell, that carries with it more dreadful devastations.

This work will therefore treat only of the pleasing application of fire, and we shall endeavor to give plain and efficient rules for the *safe* management of that element, and for the making, by means of gunpowder and other inflammable substances, various compositions, agreeable to the eye, both by their form and splendor, and to describe every principal article and instrument made use of in these amusing operations.

On the other hand, no pretension will be made to set forth an *original* set of rules and receipts, for those who

THE
PYROTECHNIST'S COMPANION.

SECTION I.

OF GUNPOWDER.

Before entering into the practice of pyrotechny, it is deemed necessary to give an ample description of the materials made use of in the art; for it must not be taken for granted that all readers are *chemists*, or that they are sufficiently versed in the science of chemistry to render such description unnecessary. But before the principles of the art can be well understood, or successfully applied, it is proper that the artist should possess a portion of *chemical* and *mechanical* knowledge; the first will teach him to select his materials with judgment, to free them from impurities, and combine them in the proportions most suitable for each particular purpose; and the latter will assist him in constructing his different pieces so as to produce the desired effect with the least loss of time and force. The description of the

CONTENTS.

SECTION IV.

DIVISION OF FIREWORKS.

SECTION V.

SINGLE FIREWORKS.

SECTION VI.

ROCKETS.

SECTION VII.

TABLES OF VARIOUS COMPOSITIONS.

term themselves *Artists in Fireworks*, whose exclusive business it is to manufacture the different articles on which it treats; to those, it is expected it will yield but little instruction; but, to the sciolistic tyro in the art, it is intended to be a *Manual of Pyrotechny*, and to treat of fireworks as objects of rational amusement; to describe, in a perspicuous manner, the materials and apparatus made use of in their construction; and to select such examples of their particular combinations, as are calculated rather for private diversion than public exhibition. The directions herein given (if strictly attended to) will enable youth to gratify their taste for this species of recreation at a comparatively small expense, and at the same time will guard them against those accidents which often arise to the ignorant, in firing the larger works purchased from the makers; and it will, throughout, strictly observe a principle of economy, the neglect of which has so frequently retarded the operations of genius.

The knowledge respecting the origin of Pyrotechny is very limited. The Chinese are said to have been the first people who had any practical acquaintance with it, or who brought the art to any degree of perfection; with them the use of fireworks is said to have been very general, long before they were known in European countries; and from accounts given of some exhibitions at Pekin, it would seem that they have attained to a degree of perfection not surpassed by any of our modern artists. Mr. Barrow, in his "Travels in China," gives, from the Journal of Lord Macartney, the following description of one of their exhibitions: "The fireworks, in some particulars," says he, "exceeded anything of the kind I had ever seen. In grandeur, magnificence, and variety, they were, I own, inferior to the Chinese fireworks we had seen at Batavia, but infinitely superior in point of novelty, neatness, and ingenuity of contrivance. One piece of machinery I greatly admired: a green chest, five feet square, was hoisted up by a pulley fifty or sixty feet from the ground, the bottom of which was so contrived as then suddenly to fall out, and make way for twenty or thirty strings of lanterns, inclosed in a box, to descend from it, unfolding themselves from one another by degrees, so as, at last, to form a collection of full five hundred, each having a light of a beautifully colored flame burning brightly within it. This devolution and development of lanterns were several times repeated, and at every time exhibiting a difference of color and figure. On each side was a correspondence of smaller boxes, which opened in like manner as the other, and let down an immense network of fire, with divisions, and compartments of various forms and dimensions, round and square, hexagons, octagons, &c., which shone like the brightest burnished copper, and flashed like prismatic lightnings, with every impulse of the wind. The whole concluded with a volcano, or general explosion and discharge of suns and stars, squibs, crackers, rockets, and grenadoes, which involved the gardens for an hour in a cloud of intolerable smoke. The diversity of color, with which the Chinese have the secret of clothing their fire, seems one of the chief merits of their Pyrotechny;" and which alone would set them upon an equal footing with the Europeans. With them, no doubt, originated the discovery of that beautiful composition, which is still known by the name of the "Chinese fire;" and the method of representing with fire that pleasing and perpetual variety of figures, which (when judiciously arranged) seem to emulate in splendor those endless beauties which adorn our celestial hemisphere. In Europe, the Florentines are said to have been the first people that gained a knowledge of the invention, and there is reason to think it was not long after the discovery of the use of gunpowder and firearms, about the end of the thirteenth, or beginning of the fourteenth century. We say the *use* of gunpowder, or application of it to firearms, for the discovery of it is believed to be of much earlier date than that generally given to it; and, whether the invention of the art of fireworks is not coeval with that of gunpowder, is a question not overburdened with improbability. The French have published several treatises on Pyrotechny, such as the *Traité des Feux d'Artifice pour la Spectacle et pour la Guerre*, by Perrinet d'Orval; the *Manuel d'Artificier*, by Father d'Incarvill, and several others of the like nature; in some of which they attach to the Chinese a *very* early knowledge of the art, and consequently the composition of gunpowder, or at least that the effects of a similar combination were not entirely unknown to them. But as the French gained their knowledge of the art from the Italians, they may probably be in an error respecting its invention;

whether they are or not is but of little importance as relating to the purpose of this work. Tracing its progress in England, it will be endeavored to give as good a delineation of the state in which it now exists as the nature of the work will admit; supposing it to be much nearer perfection than when in its earlier stages; for it is an undoubted fact that the English import nothing but what they improve.

An art which furnishes such an extensive field for amusement, reduced to plain and simple rules, digested in a familiar manner (which the most limited capacities will be able to understand), cannot fail to be entertaining to every admirer of scientific recreation.

It has been regretted by many that no publication of a like nature is now extant; and a celebrated writer, long known to the popular reader, has even said that "the English have no respectable work on the subject."

It is not presumed that the present will entirely supply such a desideratum, but to what extent it may be left for the candid reader and practitioner to determine. The author would wish it to be understood that, although he has conducted some part of the work upon mathematical principles, it is not intended as a perfect philosophical work on the subject, but as an attempt to embody into a small volume all that has hitherto been written on the subject; and if the pyrotechnic tyro receive from it any assistance towards the attainment of an art which has for its object such an endless source of entertainment, the purpose of the publication will be positively realized.

Few spectacles are more beautiful or more calculated for entertainment, than a well-conducted display of fire-works, in which such various bodies, so brilliantly illuminated, are exhibited and arranged in the most variegated forms; sometimes producing surprising and unexpected emanations, moving with velocity through the air, throwing out innumerable sparks or blazing balls, which fly off into the infinity of space; others suddenly exploding, scatter abroad luminous fragments of fire, which are trajected with the most speedy trepidation; and again, others are revolving on a quiescent centre, and by their revolutions produce the most beautiful circles of fire, which seem to vie with each other in their emanations of splendor and light.

But a faint delineation is here given of the various effects which are producible by fire, and for which we shall endeavor to give every requisite instruction for the preparation of the most pleasing garbs in which this element may be presented.

mechanical apparatus will be deferred till they come immediately under hand, and such protraction will, perhaps, be conducive to a better understanding of their utility; and, in some other section, the artist will be brought to calculate the direction which the flying fireworks (from their principles of construction) are to move, and the velocity with which they are to proceed.

Gunpowder is the principal ingredient made use of in pyrotechny; and, being of itself a compound, it will be the first object of description, and the cause of every property it possesses will be pointed out.

The invention of it is ascribed, by Polydore Virgil, to a chemist, who accidentally put some of the composition, viz., nitre, sulphur, and charcoal into a mortar, and covered it with a stone, when it happened to take fire, and (what was a natural though unexpected consequence of such combination), it shattered the stone to pieces.

Thevet says, the person here spoken of was a monk of Fribourg, named Constantine Anelzen; but Belleforet, and other authors, with more probability, suppose him to be Bartholdus Schwartz, or the Black, who discovered it, as some say, about the year 1320. The first use of it is ascribed to the Venetians in the year 1380, during the war with the Genoese; and it is said to have first been employed in a place anciently called Fossa Clodia, now Chioggia, against Lawrence de Medicis; and that all Italy made complaints against it, as a manifest contravention of fair warfare.

But this account is contradicted, and gunpowder is shown to be of an earlier era; for the Moors, when they were besieged in 1343 by Alphonsus XI., King of Castile, are said to have discharged a sort of iron mortars upon them, which made a noise like thunder. This assertion is seconded by what Don Pedro, Bishop of Leon, relates of King Alphonsus, who reduced Toledo, viz., "that in a sea-combat between the King of Tunis, and the Moorish King of Seville, about four hundred and fifty years ago, those of Tunis had certain iron tubes or barrels, wherewith they threw thunderbolts of fire."

Further, it appears that our Roger Bacon knew of gunpowder near a hundred years before Schwartz was born. That excellent friar tells us, in his treatise, *De Secretus Operibus Artis & Naturæ, & de Nullitate Magiæ*, that from saltpetre, and other ingredients, we are able to make a fire that shall burn at what distance we please; and the writer of the life of Friar Bacon says, that Bacon himself has divulged the secret of this composition in a cipher, by transposing the letters of the two words in chap. xi. of the above-cited treatise, where it is thus expressed: "Sed tamen petræ *lura mope canubre* (i. e. carbonum pulvere), et sulphuris; et sic facies tonitrum & corruscationem, si scias artificium;" and from hence Bacon's biographer apprehends the words *carbonum pulvere* were transferred to the sixth chapter of Dr. Longbain's MS. In this same chapter, Bacon expressly says, that sounds like thunder, and coruscations, may be formed in the air, much more horrible than those that happen naturally. He adds, that there are many ways of doing this, by which a city or an army might be destroyed; and he supposed that, by an artifice of this kind, Gideon defeated the Midianites with only three hundred men (Judges, chap. vii). There is only another passage to the same purpose, in his treatise *De Scientia Experimentalia;* see Dr. Jebb's edition of the *Opus Magus*, p. 474. Mr. Robins apprehends (see the preface to his tracts) that Bacon describes gunpowder, not as a new composition first proposed by himself, but as the application of an old one to military purposes, and that it was known long before his time.

Dr. Jebb, in his preface to the above-cited work, describes two kinds of fireworks; one for flying, inclosed in a case or cartouche, made long and slender, and filled with the composition closely rammed, like our modern rocket, and the other thick and short, strongly tied at both ends, and half filled, resembling our cracker; and the composition which he prescribes for both is two pounds of charcoal, one pound of sulphur, and six pounds of saltpetre, well powdered and mixed together in a stone mortar.

Mr. Dutens, in his *Inquiry into the Origin of the Discoveries attributed to the Moderns*, carries the antiquity of gunpowder much higher; and refers to the accounts given by Virgil, Hyginus, Eustathius, Valerius Flaccus, and many other writers of the same date.

To close this tedious detail, only one more work shall be mentioned, which seems to confirm the antiquity of this composition, viz., the *Code of Gentoo Laws*, 1776; in the preface to which it is asserted that gunpowder was known to the inhabitants of Hindostan, far beyond all periods of investigation.

Thus much having been said concerning the history and antiquity of this wonderful composition, it remains only to describe the method by which it is now manufactured; but to maintain that *gradatum*, or progressive order, laid down at the commencement of the work, it is necessary first to describe the ingredients of which it is composed; for it is only by a knowledge of the parts of any composition, that a good understanding can be gained of the properties of the whole.

There are only three ingredients that enter into the composition of gunpowder; these are saltpetre, sulphur, and charcoal. The first is a combination of nitric acid* and potash,† and is better known in modern chemistry

---

* *Nitric acid* is a compound of azote, or impure air, and oxygen, or vital air, and is sometimes made by repeatedly passing electric shocks through a mixture of oxygenous and azotic gas.

† *Potash*, or the vegetable alkali, is generally obtained from wood-ashes; but sometimes from the tartar, or from the lees of wine. What is used in England is generally imported from the north, where there is an abundance of wood to allow of its being burnt for this purpose.

This nitrate of potash exists in a natural state, but is generally in very small quantities. It is found at the surface of the ground in some parts of Persia and the East Indies, and is mostly united with a kind of yellowish marl, which they dig from the cliffs on the sides of hills exposed to the northern and eastern winds.

by the name of nitrate of potash. The second is a substance very well known, from the inflammable properties it possesses; it is found alone, or combined with other bodies, in various situations. In volcanic productions, it is found almost in its last degree of purity. It is found also in the state of sulphuric acid; that is to say, combined with oxygen. It is found in this state in argil,* gypsum,† &c., and it may likewise be extracted from vegetable substances and animal matter. The third and last is an article so well known in commerce, that it is almost needless to describe it—that is, charcoal; and the best to be found for the composition of gunpowder is that made from the alder, willow, or black dogwood.

This powerful composition is a mixture of these three ingredients, combined in the following proportions: for each 100 parts of gunpowder, saltpetre 75 parts, sulphur 10, and charcoal 15. In some countries, the proportions are somewhat different; but this is the combination made use of by most of the English manufacturers.

The saltpetre is either that imported from the East Indies, or that which has been extracted from damaged gunpowder. It is refined by solution, filtration, evapo-

---

* *Alumine*, or *clay;* it is found of various degrees of purity, and mixed with a variety of other earths.

† This is lime combined with sulphuric acid; it is called gypsum, plaster of Paris, plaster-stone, or selenite. It is very abundant in some parts of England, and the hills near Paris are chiefly composed of it.

ration, and crystallization; after which it is fused, taking care that too much heat is not employed, or there is danger of decomposing the nitre.

The sulphur used is that which is imported from Sicily, and is refined by melting and skimming; the most impure is refined by sublimation.

The charcoal is made in the following manner: The wood is first cut into pieces of about nine inches in length, and put into an iron cylinder placed horizontally. The front aperture of the cylinder is then closely stopped; at the other end there are pipes connected with casks. Fire being made under the cylinder, the pyroligneous acid* comes over. The gas escapes, and the acid liquor is collected in the casks; the fire is kept up till no more gas or liquid comes over, and the carbon† remains in the cylinder.

---

* From *pyr*, fire, and *lignum*, wood; the acid obtained from the partial combustion of wood. This acid is used in calico printing *as mordants* for dark-colored patterns.

† The name given by chemists to the pure part of charcoal. It is said to be present in almost all combustible bodies, and is of itself entirely of that nature. When charcoal is burnt, its carbon unites with the oxygen of the air, and so much heat as to give it a gaseous form, and constitutes carbonic acid gas, or fixed air. The same gas is also obtained by the combustion of the diamond, proving that this precious and costly article is carbon or charcoal, in a very indurated state, and assuming a determinate form. It was not till lately that the diamond was proved to be combustible; but by means of the blowpipe, and a stream of oxygen gas, it may be, to speak in common language, wholly consumed. The air that is extricated during the

The three ingredients being properly prepared are ready for manufacturing. They are first separately ground into a fine powder, then mixed in the proper proportions, and afterwards committed to the mill for the purpose of incorporating their component parts. The powder-mill is a slight wooden building, with a boarded roof, so that, in case of accidental explosions, the roof may fly off without difficulty, and in the least injurious direction, and thus be the means of preserving the other parts of the building.

The operative parts of the mill consist of two stones placed vertically, and running on another placed horizontally, which is called the bed-stone, or trough. On this bed-stone, about forty or fifty pounds of the composition are spread out, and moistened with water till reduced to about the consistency of a very stiff paste. After the stone-runners have made the proper revolutions over it, which requires about eight hours continued action of the mill, which is worked sometimes by horses, and sometimes by water, it is then taken from the mill, and sent to the corning-house, to be corned or grained. Here it is formed into hard lumps, and these are put

---

combustion is carbonic acid gas, proving the diamond to have been chiefly, if not wholly, composed of carbon.

Long before this fact respecting the diamond was ascertained, Sir Isaac Newton, reasoning from its great refracting power, declared it to be his opinion that it was one of the most combustible of bodies. Modern discoveries have now proved the fact; and it affords us an admirable instance of the acumen of that great philosopher.—*Popular Chemical Essays.*

into circular sieves, with parchment bottoms, perforated with holes of different sizes, and fixed in a frame connected with a horizontal wheel. Each of these sieves is also furnished with a runner or spheroid of lignum vitæ, which, being set in motion by the action of the wheels, forces the paste through the holes of the parchment bottom, forming grains of different sizes. The grains are then separated from the dust by sieves and reels made for that purpose. The grains are next hardened, and the rougher edges are taken off by shaking them for some time in a close reel, moved in a circular direction with a proper velocity.

When the powder has been corned, dusted, and glazed, it is dried in the stove-house, where great care should be taken to avoid explosion. The stove-house is a square apartment, three sides of which are furnished with shelves or cases, on proper supports, arranged round the room; and the fourth contains a large cast-iron vessel, called a "gloom," which projects into the room, and is heated from the outside, so that no part of the fuel may touch the powder. For greater security against sparks by accidental friction, the glooms are covered with sheet-copper, and are always cool when the powder is put in or taken out of the room.

Here the grains are thoroughly dried, losing in the process what remains of the water added to the mixture in the mill for bringing it to a working stiffness. A method of drying powder by steam-pipes, running round and crossing the apartment, has been successfully tried; and thus the possibility of any injurious accident from over-heating is prevented. The temperature of the room, when heated in the common way by a gloom-stove, is always regulated by a thermometer hung in the door of the stoves.

If gunpowder is injured by damp in a small degree, it may be recovered by again drying it in a stove; but if the ingredients are decomposed, the nitre must be extracted by boiling, filtering, evaporating, crystallizing, &c., and then, with fresh sulphur and charcoal, be re-manufactured.

There are several methods of proving and trying the goodness and strength of gunpowder. The following, as common methods, are frequently made use of: 1. By sight; for, if it be too black, it is too moist, or has too much charcoal in it; and if rubbed upon white paper, it blackens it more than good powder does. 2. By touch; for if, in crushing with your finger-ends, the grains break easy, and turn into dust, without feeling hard, it has too much charcoal in it; or if, in pressing under your fingers upon a smooth, hard board, some grains feel harder than the rest, or, as it were, dent your finger-ends, the sulphur is not well mixed with the nitre, and the powder is bad. And also by burning, in which method, little heaps of powder are laid on white paper three or four inches asunder, and one of them fired; which, if the flame ascend rapidly, and with a good report, leaving the paper free from white spots, and without burning holes in it, and if sparks fly off and set fire to the adjoining heaps, the quality of the powder may be safely relied on; but if otherwise, it is either badly made or the ingredients are impure.

These are some from among the common methods made use of for this purpose; but for greater accuracy in determining the relative strength of gunpowder, various machines have of late been invented by men connected with military affairs. That excellent mathematician and philosopher, C. Hutton, LL. D., F. R. S., and late Professor of Mathematics in the Royal Military Academy, Woolwich, has constructed a machine for this purpose, which, for convenience and accuracy, far surpasses anything of the kind hitherto invented. It is called Eprouvette, or a Gunpowder Prover (for plans and description, see third vol. *Hutton's Tracts*, page 153); and from its possessing so many peculiar advantages, is now generally used. It consists of a small cannon, the bore of which is about one inch in diameter, suspended freely like a pendulum, with the axis in a horizontal direction. This being charged with the proper quantity of powder, which is usually about two ounces, and then fired, the gun swings or recoils backward, and the instrument itself shows the extent of the first or greatest vibration, which indicates the strength to the utmost nicety. The whole machine is so simple, easy, and expeditious in its use, that the weighing of the powder is the greatest part of the trouble; and it is also so uniform with itself, that the successive repetitions or firings with the same quantity of the same kind of powder, hardly ever yield a difference of the hundredth part from the first vibration.

Having thus given an account of almost everything necessary to be known in regard to the process of making and ascertaining the relative strength of gunpowder, this article shall be closed with a few observations (which will be selected from the best authorities) on the physical causes of its inflammation and exploding. When the several ingredients of gunpowder are properly prepared, mixed, and grained, in the manner already described, if the least spark be struck thereon from a steel and flint, the whole will be immediately inflamed, and burst out with extreme violence.

The effect is not hard to account for; the charcoal part of the grains whereon the spark falls, catching fire like tinder; the sulphur and the nitre are readily melted, and the former also breaks into flame; and at the same time the contiguous grains undergo the same fate. Now it is known that saltpetre, when ignited, rarefies to a prodigious degree. Sir Isaac Newton reasons thus on the subject: "Charcoal and sulphur in gunpowder easily take fire, and kindle the nitre; and the spirit of the nitre, being thereby rarefied in the vapor, rushes out with an explosion much after the manner that the vapor of water rushes out of an æolipils; the sulphur also, being volatile, is converted into vapor, and augments the explosion: add, that the acid vapor of the sulphur, namely, that which distils under a bell into oil of sulphur, entering violently into the fixed body of the nitre, lets loose the spirit of the nitre, and excites a greater fermentation, whereby the heat is farther augmented, and the fixed body of the nitre is also rarefied into fume; and the explosion is thereby made more vehement and quick."

If salt of tartar be mixed with gunpowder, and that mixture be warmed till it take fire, the explosion will be greatly more violent and quick than that of gunpowder alone; which cannot proceed from any other cause than the action of the vapor of gunpowder upon the salt of tartar, whereby the salt is rarefied.

The explosion of gunpowder arises, therefore, from the violent action whereby all the mixture, being quickly and vehemently heated, is rarefied and converted into fume and vapor; which vapor, by the violence of that action, becomes so hot as to shine, and appear in the form of a flame.

Another cause of the effects of gunpowder may be owing to the sudden formation of a quantity of gas, and are consequently greater when the gas is confined in all directions but one, as in our guns and cannon. The nitric acid of saltpetre is decomposed, and affords the gas. The other ingredients dispose it to be easily inflamed, which is necessary to the decomposition of the acid. Dr. Ingenhousy accounts for the effect of gunpowder by observing that nitre yields by heat a surprising quantity of pure dephlogisticated air, and charcoal a considerable quantity of inflammable air; the fire employed to inflame the powder extricates these two airs, and sets fire to them at the instant of their extrication.

Count Rumford is of opinion that the force of the elastic fluid, generated in the combustion of gunpowder, may be satisfactorily accounted for upon the supposition that its force depends solely on the elasticity of watery vapor or steam.

M. De la Hire, in the history of the French Academy for 1702, ascribes all the force and effect of gunpowder to the spring or elasticity of the air inclosed in the several grains thereof, and in the intervals or spaces between the grains; the powder being kindled, sets the springs of so many little parcels of air playing, and dilates them all at once, whence the effect; the powder itself only serving to light a fire which may put the air in action, after which the whole is done by the air alone.

Dr. Hutton seems to differ from the opinion of M. De la Hire, in regard to the expansion of inflamed gunpowder. Is it, he observes, occasioned by the air interposed between its grains, or by the aqueous fluid which enters into the composition of the nitre? We doubt much (continues he) whether it be the air, as its expansibility does not seem sufficient to explain the phenomenon; but we know that water, when converted into vapor by the contact of heat, occupies a space 14,000 times greater than its original bulk, and that its force is very considerable.

The same learned author says that the discovery of the true cause of the expansive force of fired gunpowder is chiefly due to the English philosophers, and particularly to the learned and ingenious Mr. Robins. This author apprehends that the force of fired gunpowder consists in the action of a permanently elastic fluid, suddenly disengaged from the powder by the combustion, similar in some respects to common atmospheric air, at least as to its elasticity. He showed, by satisfactory experiments, that a fluid of this kind is actually disengaged by firing the powder; and that it is *permanently* elastic, or retains its elasticity when cold, the force of which he measured in this state. He also measured the force of it, when inflamed, by a most ingenious method, and found its strength in that state to be about a thousand times the strength or elasticity of common atmospheric air. This, the Doctor observes, is not its utmost degree of strength, as it is found to increase in its force when fired in larger quantities than those employed by Mr. Robins; so much so, indeed, that, by more accurate experiments, we have found its force rise as high as 1600 or 1800 times the force of atmospheric air in its usual state. Much beyond this it is not probable it can go, nor indeed possible, if there be any truth in the common and allowed physical principles of mechanics. With an elastic fluid of a given force, we infallibly know or compute the effects it can produce in impelling a given body; and on the other hand, from the effects or velocities with which given bodies are impelled by an elastic fluid, we certainly know the force or strength of that fluid, and these effects have been found perfectly to accord with the force above mentioned. Mr. Robins's discovery and opinions have also been corroborated by others, among the best chemists and philosophers. Lavoisier was of opinion that the force of fired gunpowder depends, in a great measure, on the expansive force of uncombined caloric, supposed to be let loose in a great abundance, during the combustion or deflagration of the powder. And Bouillon Lagrange, in his course of Chemistry, says, when gunpowder takes fire there is a disengagement of azotic gas, which expands in an astonishing manner when set at liberty; and we are even still ignorant of the extent of the dilatation occasioned by the heat arising from the combustion. A decomposition of water also takes place, and hydrogen gas is disengaged with elasticity; and by this decomposition of water there is formed carbonic acid gas, and even sulphuretted hydrogen gas, which is the cause of the hepatic smell emitted by burnt powder.

It has been found by experiment, that granulated powder inflames with much greater rapidity than that which is not granulated; the latter only puffs away slowly, while the other takes fire almost instantaneously; and of the granulated kinds, that in round grains much sooner than that in oblong irregular grains; the cause of which may arise from the former leaving to the flame larger and freer interstices, which produce the inflammation with much more rapidity.

Gunpowder is supposed to explode at about 600° Fahr., but if heated to a degree just below that of faint redness, the sulphur will mostly burn off, leaving the nitre and charcoal unaltered.

Experiments have also proved that the variations in the state of the atmosphere do not in any way alter the action of powder. By comparing several trials made at noon in the hottest summer sun, with those made in the morning and evening, no certain difference could be perceived; and it was the same with those made in the night, and in winter. And indeed, considering the principles of the explosion, and that it always contains the same quantity of the elastic fluid, it is difficult to conceive how its force can be affected by the density or rarity of the atmosphere.

The action and nature of this formidable composition being now somewhat fully described, the principal object of the work will now be proceeded with—that of constructing the most common and curious articles for pyrotechnic exhibitions.

SECTION II.

MATERIALS.

Having, in the preceding section, entered somewhat largely on the nature and properties of gunpowder, and consequently of the ingredients which compose it, any further observations on them would be unnecessary, providing the ingredients and proportions always remained the same. But as the ingredients used in the manufacture of that article are frequently employed in various other proportions, to form compositions for filling fireworks, it is necessary to give some further directions for the choice and purification of these articles, which, together with the apparatus used in the making of fireworks, will form the subject of the present section.

1. Nitre.—Among the various articles made use of in the composition, none are of greater importance than saltpetre; for on the quantity and purity of this, depend all the force and much of the brilliancy of the fire. The most common sort is that usually sold by the grocers, and is generally in large lumps formed of an assemblage of small crystals somewhat transparent, and often

mixed with earthy matter and many other impurities. In its purest state it is in the form of small, six-sided, prismatic crystals, not apt to grow moist or powdery on exposure to the air. The pure nitre is now become very expensive, so that it is of consequence to know how the common nitre, or nitre of commerce, may be purified, for it is found to answer no purpose in pyrotechny unless such change or purification has been effected.

Nitre is found (like most of other saline bodies) to be much more soluble in boiling water, than in water of the ordinary temperature. If, therefore, the nitre of commerce be dissolved in a small quantity of boiling water, and the solution be properly strained, the liquor, when cold, will afford crystals that are very pure. The following is the most convenient method of proceeding: Dissolve the nitre in boiling water (which should be soft water), in the proportion of about a quart to each pound of nitre; and that the solution may be more easily effected, let the nitre be reduced to a powder previous to its being immersed, and let the vessel containing the nitre and water be kept at the boiling heat till all the salt is dissolved; then strain the liquor, while hot, through thick blotting-paper, placed in a clean funnel; and set by the filtered liquor in a shallow vessel, in a cold place, to crystallize. The crystals thus obtained are to be dried, first in blotting-paper, and then before the fire, and kept for use. From the remaining solution, which is sometimes called *mother-water*, fresh crystals may be procured by boiling it in a clean tin vessel till a filming scum arises to the surface, then filtering it through paper, and setting it to crystallize as before.

Very pure nitre may also be obtained from damaged gunpowder, which may be sometimes procured at a cheap rate, at the shops where it is sold for this purpose. The damaged powder must be ground with a small quantity of hot water, in a large wooden or stone mortar, or it may be boiled over a gentle fire, with sufficient water to cover it, till as much as possible of the nitre is dissolved; the liquor is then to be strained through a thick flannel bag, afterwards filtered while hot through blotting-paper, the sediment to be boiled down till a film rises on the surface; again filtered and set by to cool and crystallize, as directed in the process for the former method.

The nitre must always be reduced to fine powder, previous to mixing it with other substances; this is easily done by dissolving it in a little more than its own weight of boiling water, keeping the solution over a gentle fire, and continually stirring it with a flat stick till all the water is evaporated, when the powder is to be taken out and dried before a gentle fire; during which, care must be taken not to let it remain too long, or exposed to too great a heat, otherwise it will be melted into a firm cake. The drying may be completed by suffering it to remain a sufficient time on paper before the fire. For the purification of saltpetre, both these methods may (by attending to the foregoing instructions) be practised with success; but of the two, the former is more strongly recommended.

2. Sulphur.—Sulphur is the next ingredient, in regard to importance, as being the most inflammable material known. It exists in three states, in all of which it is occasionally employed in fireworks; the first is that brought from the neighborhood of volcanoes, and is called *native sulphur*, but more commonly *sulphur vivum*, though (it may be observed), what is sold in the shops under this name is a drossy powder, the refuse left after purification. The second is that in the roll, called *roll sulphur*, or *stone brimstone*. The third is the *sublimated sulphur*, or, as it is commonly called, *flower of sulphur*; this, when genuine, is the purest, and is found to answer best for all nice and delicate articles, and from its being already in a state of powder it is by far the most convenient, as the others require to be ground or mealed previous to their being used. The first kind is the cheapest, and answers pretty well for all large and coarse articles, but as it is most frequently mixed with earthy matter and other impurities, the use of it cannot be much recommended. The second is found to be the strongest, and which is mostly used, particularly for most of the ordinary articles; but such is the desire of gain, that this article of sulphur is not suffered to pass through the hands of the dealers without its quality being reduced by adulteration, which they effect by mixing with it resin, flour, &c.; when pure, it is of a bright yellow color, dense but not too heavy, easily cracks with the heat of the hand, and the broken parts look bright and crystallized. There is another kind of *sulphur* (though not generally known

among dealers) which does not burn like others, and, what is rather singular, it emits no sulphurous smell, for being put upon the fire it melts just like common wax. This sort is found in great abundance in Iceland, near Mount Hecla, and Carniola. This sulphur is commonly of a reddish color, like that found in the Straits of Heildesheim, where it is likewise of several colors, as pale yellow and green, and generally adheres to the surface of stone and rocks, from which it may be easily broken off and collected; that which is perfectly yellow of each kind is the best. That of the first description, or *sulphur vivum*, is sometimes called *quick sulphur*, from its undergoing no change by fire, since its production by nature; and in some countries it is called *virgin sulphur*, because the women and girls in Campania frequently make a kind of paint of it, for a no less delicate purpose than that of beautifying the face. Should either kind be met with in an impure state, the following method may be applied for the purpose of purification.

To purify Sulphur.—Melt a quantity of it in an iron pan, by which means the earthy and metallic parts will be precipitated, and then pour it into a copper-kettle, where it will form another deposit of the impure matter with which it is mixed; after keeping it for some time in a melted state, pour it into cylindric wooden moulds, for the purpose of forming it into sticks; the moulds may be about an inch in diameter; their length may be various. If the sulphur should take fire during this operation, it may be quickly extinguished by covering the pan close over at the top.

3. CHARCOAL is also a considerable ingredient in these compositions, but is of a much more simple nature than that of the foregoing ones. It may generally be procured at the hardware shops, or at foundries, or it may be easily prepared; for which, put a quantity of small pieces of wood, such as beech or alder, into a large earthen or iron pot, filling up the vacuities, and covering the top with sand; then, placing the pot in the middle of a strong fire, and keeping it at a red heat for two or three hours, as the sand excludes the air, the wood is reduced to charcoal without the possibility of its being consumed; and when the pot is cold, the charcoal is to be taken out and kept for use in some very dry place. Small quantities should only be made at a time, as it is always best newly prepared.

4. STEEL-DUST is another important ingredient in fire-works, for being mixed with mealed powder or some other composition, and the mixture inflamed in a proper tube, or case, the jet of fire produces a most brilliant appearance by the sparks arising from the ignition of the iron in the oxygen gas of the nitre.

Iron-filings (for this steel-dust is nothing more than pure iron reduced into small particles by filing or some other method), when free from rust, and not mixed with any impurities, answer very well; but firework-makers generally prefer cast-iron reduced to powder by beating thin plates of it on a cast-iron anvil with a heavy hammer, and sifting the broken particles through sieves of brass or iron wire, of different degrees of fineness, so as to separate the particles into grains of various sizes, according to the magnitude of the pieces. The grains thus sorted have been called *iron-sand*, and have been distinguished into sand of three or four *orders*, according to their respective fineness; thus the sand that passes through the first sieve is called sand of the *first order;* and that which passes through the second, sand of the *second order;* and so on to the fourth, which is generally very coarse. The finest is calculated for fireworks of the smallest size, the second for pieces somewhat larger, and that of the last order, only for pieces of the largest size, such as gerbes of six or eight pounds, the composition of which being of proportionate strength to bring such large particles into a state of ignition.

As these grains are very apt to rust by keeping, they should be preserved either in close-stopped bottles, well dried, or in boxes that shut close, and are lined with paper moistened in linseed oil. It sometimes happens that fireworks may be required to be kept a long time, or sent abroad; which could not be done with the brilliant fires, if made with filings unprepared, for this reason, that the saltpetre being of a damp nature causes the iron to rust; the consequence of which is that, when the works are fired, there will appear only very few brilliant sparks, but instead of them a number of red and drossy sparks; and besides, the charge will be so much weakened that if this were to take place in wheels, the fire would scarcely be strong enough to force them round; but to prevent such failures in the firing of them, the filings, or iron-sand, may be thus prepared :—

To prepare iron-sand.—Melt some brimstone in a glazed earthen pan, over a slow fire, and when dissolved throw in some filings, which keep stirring till they are all covered with brimstone; this must be done while it is on the fire; then take it off and stir it very quickly till cold, when it must be rolled on a board with a wooden roller, till broken as fine as corned powder, after which sift from it as much of the brimstone as possible. In heating the brimstone, should it take fire, it may be quickly extinguished by covering the pan close over at the top.

Second method.—There is another method of preserving filings so as to keep two or three months in winter; this is done by rubbing them between strong brown paper, which has been previously moistened with linseed oil.

In fine, it will be well to anticipate a little trouble in the preparation of this granulated iron-sand; for cast-iron being of so hard a nature as not to be cut by a file, it is obliged to be pulverized, or reduced to grains by the method described, which is rather difficult to perform; but when it is considered what beautiful sparks this iron yields, no pains should be spared to granulate such an essential material.

When these plates of iron cannot be procured, an old cast-iron pot may be employed; but especial care must be taken that its surface be perfectly free from rust and other impurities, previous to its being pulverized, otherwise it will entirely destroy the effect it is intended to produce.

It is to the Chinese we are indebted for this method of rendering fire so brilliant and variegated in its colors, who discovered it long before Father d'Incarville made it known to the European countries. This sand, when it inflames, emits a light exceedingly vivid; and it is surprising to see fragments of this matter not larger than a poppy-seed form all of a sudden luminous flowers of stars, twelve and fifteen lines in diameter. These flowers are also of different forms, according to that of the inflamed grain, and even of different colors, according to the matters with which the grains are mixed. But rockets, into which this composition enters, cannot be long preserved, unless prepared as described in the early part of this article.

There are many other substances occasionally employed in the composition of fireworks, but as they can be procured at all chemists and druggists ready for the purpose, it is unnecessary to give any detail respecting them, further than enumeration. They are chiefly the following, viz. *camphor*, which is used to improve the appearance of the fire; *antimony*, or *sulphuret of antimony*, *sal-ammoniac*, *verdigris*, and *pitch*, for giving to the fire different and particular shades of color; *flowers of benjamin*, or *benzoic acid*, for imparting to it an agreeable odor; and *spirits of wine*, or *camphorated spirits*, for mixing up the ingredients into a paste. These liquids are found to answer much better than common water, or gum-water, which is sometimes used, as they do not dissolve the saltpetre, and are therefore not so liable to produce a separation of the materials employed. *Lamp-black* is sometimes used instead of charcoal, and is supposed to have the effect of diminishing the heat of the fire, while it does not materially lessen its brilliancy. Thence it is a considerable ingredient in what is called *cold-fire*, the seeming paradox of which will be hereafter reconciled. For the same purpose, that of diminishing the force of the composition, *powdered glass* and *sawdust* have been frequently employed; but probably these effects might be better, and with more certainty answered, by lessening the proportion of nitre.

5. Oil of Camphor.—This liquid is frequently used for the purpose of moistening the compositions; it is thus easily procured: Put a small quantity of camphor into a brass mortar, and add to it some oil of sweet almonds, sufficient to reduce it to a stiff paste, then work the mixture well together, and it will turn to a green color, after which add a sufficient quantity of oil to liquefy it for use. It must be observed, on the use of this liquid, that the composition into which it enters must be kept as much as possible from the air, as an exposure to it will cause it to evaporate, and thereby occasion a failure in the exhibition.

6. Benzoin.—Benzoin, or as it is vulgarly called Benjamin, is a resinous matter obtained from the tree called *Benzoin*, and is brought from different parts of the Indies, where it is found of various kinds, and of different colors; the best is that which is full of white

spots, and is easily broken. It is used in odoriferous fireworks, but prior to which it must be reduced to a fine powder, which may be effected by the following method: Put about three or four ounces of benzoin grossly pounded into a deep and narrow earthen pot, and cover the pot over with a cone of thick paper, tying it closely round the edge, then place the pot over the fire and apply a moderate heat; after the interval of an hour, take off the cone, and you will find some flower sticking to the underside of it; or, in the language of chemistry, the acid is sublimated and is deposited on the paper; the cone must be returned to the pot, and the operation continued till the flower appears very white and fine.

The acid, which is frequently used, may be obtained by digesting benzoin in sulphuric acid, and by this it is obtained much purer and in finer crystals than by any other method.

On this article it has been deemed well to give the above information; but to the private practitioner, it will be more eligible to purchase it ready prepared.

SECTION III.

APPARATUS.

In the practical part of pyrotechny, the constructing and due proportioning of the moulds is a very material consideration, for on these the goodness of the article depends nearly as much as on the purity of the ingredients. They consist chiefly of solid and hollow cylinders, made either of wood or metal; those that are hollow are called *moulds*, and those that are solid *formers;* both are used in the construction of rockets; similar cylinders, either of wood or metal, are used for ramming down the composition; a machine for contracting the aperture of the cases, the operation of which is called *choking;* another for boring them after they are filled; and a simple apparatus for grinding the materials previous to the cases being filled, as well as others of less importance, which will be described as their assistance is required.

In describing the important apparatus, those which come most immediately into use, will be first noticed.

1. Grinding Machines.—For the purpose of triturating or properly mixing the several ingredients

together, various contrivances have been resorted to. A common iron mortar, such as is used by druggists and apothecaries, is found to answer very well for grinding or pounding the brimstone, charcoal, saltpetre, &c., *separately*; and apothecaries close sieves, fitted with wire-cloth, are the best possible implements for obtaining the fine powder; but when corn gunpowder is to be *mealed*, or the various ingredients are to be mixed together, such mortars cannot be used, as the heat generated by the continued action of the pestle might inflame the mixture, and thereby place the life of the operator in imminent danger. To obviate these dangerous probabilities, a very simple contrivance has been effected; this is called the mealing-table, and for the purpose has proved very speedy and effectual. It consists of a rectangular elm board, with a rim round its edge, four or five inches high, at one end of which a part of the rim is made to slide in a groove, so that, after mealing the powder, it may be swept clean out from the table. A representation of it is here given, Fig. 1. Fig. 2

Fig. 1.

Fig. 2.

Fig. 3.

is a small copper shovel, generally made use of for filling and emptying the table. When about to meal a quantity of powder, observe not to put too much on the table at once; but when a moderate portion is put on, take the muller (Fig. 3) and rub it till all the grains are well broken; then sift it in a lawn sieve, that has a receiver and top to it, such as is generally used by the apothecaries, and that which does not pass through the sieve must be returned to the table, and, with an additional quantity, ground over again. Sulphur and charcoal may be ground in the same manner, only these being much harder than powder the muller must be of ebony, or any other hard wood, else the ingredients will stick in the grain of the elm, and be very difficult to grind. As sulphur is apt to stick and clod to the table, it will be found best to have one for that purpose, as they are easily procured; this will be but little trouble, and will be more than compensated for by the sulphur being always kept clean and well ground.

The following is another method for the above purpose, which some consider equally effective. This is a mortar made of hard wood, shaped like that of the druggists, with the bottom rounded within, and having a wooden lid fitting close on the top, and in the centre a hole just large enough to admit easily the stalk of the pestle, to the lower end of which is connected a piece of marble terminating in a spherical surface. With this apparatus gunpowder may be safely ground to meal, or its ingredients mixed by the continued motion of the pestle in the hole of the lid.

Method of mixing the Ingredients.—Connected with that of grinding is the operation of mixing the ingredients, and which is considered a principal part of the business of pyrotechny; and indeed many articles depend as much on the well-mixing as on the proportion of their composition; therefore great care should be taken in this part of the work, and particularly so in the composition of skyrockets. When about four or five pounds of ingredients have been duly prepared for mixing (which is a sufficient quantity to mix at one time), first put them together in some vessel convenient for the purpose, then work them about with the hands, till their various natures are pretty well incorporated; after which put them into the lawn sieve with the receiver and top to it, and sift it into some other clean vessel, and if any remains that will not pass through the sieve, grind it again till fine enough; and if it be suffered to pass twice through the sieve, it will be more than the trouble the better. For rockets and all fixed works, from which the fire is to play regularly, the ingredients must be prepared as above; and it may be observed here, that all compositions which contain steel or iron filings must be mixed or shifted with the copper shovel, for the hands are apt to impart a moisture which is injurious to their nature. Nor will any works which have iron or steel in their charge keep long in damp weather without being properly prepared, as was directed in the preceding section.

There are several other moulds and apparatus made use of, but as most of them are used in the making of rockets, and some few other articles, and are so immediately connected with the practice thereof, their use and application will be better understood when treating of that article in the next section, rather than by giving their descriptions in this place.

SECTION IV.

DIVISION OF FIREWORKS.

FIREWORKS are generally divided into two classes: those which compose the first, are chiefly *squibs, serpents, crackers, sparks, marroons, saucissons, pin-wheels, leaders, gerbes,* or *Roman candles,* and (when without any appendages) *rockets;* these, by their requiring but little dexterity in the preparation, are called simple, or more properly, single fireworks, and are said to be of the first class. Others, which are of more difficult construction, are called compound or complex fireworks, and are said to be of the second class. These consist of *suns, moons, stars, wheels, globes, balloons, batteries, flower-pots, fire-pumps, pyramids,* &c.; these are generally composed of some of the single pieces, as gerbes, serpents, marroons, saucissons, &c., properly arranged on suitable frames according to the taste of the operator, and connected with each other by long pipes filled with inflammable composition called leaders, and fired by means of *quickmatches,* or *port-fires,* and very frequently by common touch-paper. The descriptions and instructions will begin with those of the simple or single kind, which will lead progressively to those which are more complex, in the order proposed at the commencement of the work.

In the subsequent directions there will be frequent occasion to mention pipes of communication, commonly called *leaders*, by which the several parts of a compound work are connected with each other; and several other articles of less importance, as touch-paper, quick-match, port-fires, &c.

1. TOUCH-PAPER.—This is a paper impregnated with a solution of saltpetre, by which it acquires the property of burning slowly away without flame, and yet with sufficient strength to communicate its fire to mealed powder, when it comes in contact. It is prepared in the following manner:—

To make Touch-paper.—Dissolve a quantity of saltpetre in vinegar or any other acid, more or less of the saltpetre according as the paper is required to burn slow or fast; then dip into this solution some thin blue paper, let it be well saturated, then take it out, and dry it for use. If on trial, it is not found to burn properly, or if it blazes on being set on fire, it is an indication that the solution is too weak; it must therefore be strengthened by adding more of the nitre, and the paper must be passed through again. On the application of this paper to fireworks, two modes are in use: For small articles, or such as are *choked* (to be hereafter explained), tie a piece round the orifice with thread or fine twine, leaving enough of the paper at the end to form a small tube, in which is put some mealed gunpowder, and the paper is then twisted over it, and ready for firing.

For larger articles, as rockets, Roman candles, &c., the paper, instead of being tied, should be pasted round the orifice with thin flour-paste; but care must be taken that the paste does not extend beyond the end of the case, for this would prevent the fire from communicating with the composition, and the piece would consequently fail in the going off.

2. QUICKMATCH.—The purpose of the quickmatch is similar to that of touch-paper, but is chiefly used to form the inside of leaders; it is generally made of cotton-wick (such as is usually made use of in the manufacture of candles), impregnated with nitre. It is made of several sizes, from one to six threads, as is most suited to the pipes, or articles for which it is designed. The pipes must be sufficiently large to receive the match easily, as its quality will be much diminished by its breaking. The following is the best method of making this match: Having distributed the cottons into the number of threads requisite for the purpose, coil it very lightly into a flat-bottomed copper or earthen pan; then pour in a part of the saltpetre and liquor, and boil them together about twenty minutes, after which coil it again into another pan and put to it the remainder of the liquor; then put in some mealed powder, and well mix it with the liquid; after which place the pan beneath the wooden frame (Fig. 4), tying one end of the cotton to one side of the frame; then by one hand, by means of the handle (A), turn the frame round while the cotton

passes through the other, holding it very lightly, and at the same time keeping the hand full of the wet powder. If the powder is too wet to stick to the cotton, put more

Fig. 4.

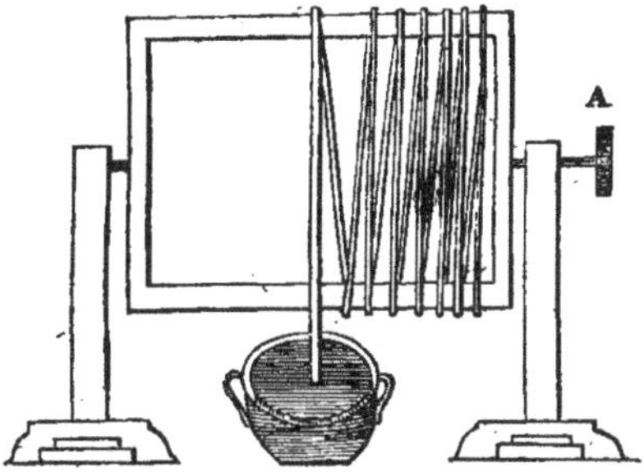

in the pan, so as to keep a supply until the match is all wound up; it may be wound as close on the frame as is thought proper, provided it does not stick together. When the frame is full, take it off the joints and sift dry mealed powder on both sides the match, till it appears quite covered; afterwards hang it in some warm place to dry, which, if it be in summer, will be effected in a few days; but if it be in winter, it will be a fortnight before it is fit for use; when it is perfectly dry, cut it along the outside of one of the side-pieces of the frame, and tie it up in skeins for use.

The proper ingredients for the match are, cotton one pound twelve ounces, saltpetre one pound, spirits of wine two quarts, water three quarts, isinglass three gills, and mealed powder ten pounds; or half the quantity may be prepared by taking the ingredients in the same proportion. Four ounces of isinglass should be dissolved in about three pints of water.

3. Port-fires.—This term is applied to paper tubes, filled with mealed powder, or a similar composition, and which is generally used in setting fire to rockets, or compound fireworks, which require to be lighted very expeditiously. There are two kinds, the one used as above, the other for illuminations: those of the former kind are usually called common port-fires, and may be made of any length, but are seldom more than twenty-one inches; they are rolled on rods about half an inch in diameter, and made of cartridge-paper in three or four folds till their exterior diameter is about five-eighths of an inch, the last fold being well secured at the edge by paste, and one end pinched or folded down. The moulds, five-eighths of an inch diameter, should be made of gun-metal, and to take to pieces lengthwise, forming two semi-cylindrical tubes, and when used, to be connected together by several rings fitted to the outside of the tube. If about an inch of metal be made fast to one extremity of the half tube of the diameter of the rod, or former, it will supersede the necessity of a foot, and be much more convenient: but the part of the former, as it may be termed, must be made very fast to the tube, or it will easily be detached by the ramming of the cases. The composition for filling these cases consists of saltpetre, sulphur, and mealed powder. In using these port-fires, the close end is fixed in a metal socket made like a port-crayon, which is attached to a

stick of sufficient length to reach any required part of the firework.

The following is the composition for all fireworks:—

Sulphur	2.0
Saltpetre	6.0
Mealed powder	1.0

For a brilliant fire add, steel filings 1.0, and for a dark fire add, charcoal 1.0.

4. PORT-FIRES FOR ILLUMINATIONS.—These differ only as regards their length from those above described; their diameter is the same; their length from three to six inches, pinched close at one end and left open at the other; they are filled by small quantities at a time, and rammed very lightly, or their cases will be endangered. Three or four rounds of paper, with the last round pasted, will be strong enough for these cases; the composition the same as before.

5. LEADERS, *or Pipes of Communication.*—These are small tubes of paper, of lengths adapted to the distances to which they are to extend, and filled with a combustible composition that will not burn too fast. As it is much the best to have them in long lengths, some large size paper must be used for the purpose; that which is called "Elephant" is found most convenient, and for this purpose is generally used. It is cut into strips two or three inches broad, or sufficient to go four times round the *formers*, which will make the tube strong enough for most ordinary purposes; indeed, if they are made with greater substance, much inconvenience will

be found in the application of them to the different works for which they are designed, from flying off without communicating their fire.

The *formers* for these leaders should be about one-fourth of an inch diameter; this size has been found to answer most purposes, though they are sometimes made of less, as well as larger diameters, but from one-eighth to three-eighths must be the extremes; smooth brass wire of proper dimensions make the best *formers* that can be used. When used, observe to dip them in oil or grease to prevent their sticking to the paper, which must be pasted all over; in rolling them, make use of a rolling-board, but press it very lightly upon them; when the former is drawn out, which must be done with one hand while the tube is retained with the other, great care must be taken, or the former will stick and tear the paper.

In the joining and placing these leaders, it is necessary to be as particular and careful as in their manufacture, for on the well securing and adjustment of them depends much of the performance of all complex pieces; on this account it will be necessary to give in detail, and in as plain a manner as possible, the best method: the works being ready to be clothed (as this operation is termed), cut the pipes in lengths sufficient to reach from one case to the other, then put in the quickmatch (prepared as taught in the last article), which must always be made to go in very easy; when the match is in the tube, cut it off about an inch beyond the end of the pipe, and let it project as much at the other end; then fasten the pipe to the mouth of each case with a pin, and put the loose ends of the match into the mouths of the cases of the works, with a little mealed powder; this done, paste over the mouth of each two or three strips of paper, and the joint will be well secured.

For illuminations and small cases, the following method is generally adopted:—

First, thread a long pipe, then lay it on the tops of the cases and cut a piece off the underside over the mouth of each case, so that the match may appear; then pin the pipe to the other cases; but, before putting on the pipes, put a little mealed powder into the mouth of each case. If the cases thus clothed are port-fires or illuminated works, cover the mouth of each case with a single paper; but if they are choked cases, so situated that a number of sparks from other works may fall on them before they are fired, secure them with three or four papers, which must be pasted on very smoothly so that there be no creases for the sparks to lodge in, which often set fire to the works before the time.

Avoid as much as possible placing the leaders too near or one across the other, so as to touch, as it may happen that the flash of one will fire the other, and thereby destroy the beauty of the arrangements.

If the works should be so formed that the leaders must cross or touch each other, be very careful to make them strong and secure at the joints, and likewise at every opening.

When a great length of pipe is required, it must be made by joining several pipes together, after the following manner: Having put on one length of match as many pipes as it will hold, paste paper over every joint; but, if a still greater length is required, more pipes must be joined by cutting about an inch off one side of each pipe near the end, and laying the quick-matches together and tying them with small twine, after which cover the joining with pasted paper.

SECTION V.

OF SINGLE FIREWORKS.

THAT class of articles shall now be enumerated and described, which, from the simplicity of their construction, have obtained the name of Single Fireworks; among these, the first which offers itself to notice is the serpent, or what is commonly called the squib.

1. SERPENTS. — The serpents are generally made about six or eight inches long, and about half an inch in diameter; they are sometimes made straight, and sometimes with a choke in the middle of them; the name which they bear, probably arose from the hissing noise which they make when fired, or from the zigzag or vibratory directions in which they move, when properly constructed, on being projected from the hand. Fig. 5 represents a serpent complete, where A C, the length of the case, may be about six inches for an ordinary size. These cases must be made of some strong paper, and rolled on a former about one-fourth of an inch diameter, or somewhat more, and having choked or tied one end up close, with strong twine, fill the case about two-thirds of the way with some of the composition described in the general table in Section VII., rammed

Fig. 5.

moderately hard in the mould proper for the diameter of the case, and then it is either choked in the part B, that is, pinched with a piece of twine, so as to leave a very small aperture, or some obstructing body, such as a small piece of paper, or a vetch-seed is introduced, and the remainder of the case must be filled with grain or corn powder. Lastly, the other extremity is to be well secured with twine, and is commonly dipped into melted pitch; the other end must be now untied, and a little moistened mealed powder is introduced, over which a piece of touch-paper being properly fastened, the serpent is complete.

If the serpents are not choked towards the middle, instead of moving in a zigzag direction, they will ascend and descend with an undulating motion, till the fire is communicated to the grained powder in the part B C, when they will burst with a loud report.

To introduce the composition into small cases, a quill cut into the form of a spoon will be found very useful. The trouble of first temporarily choking, and tying the ends of the cases, may be dispensed with, if the moulds in which they are rammed have attached to them a foot and nipple, as described in the article ROCKETS.

The common squibs, or such as are of small dimensions, may be made with still less trouble, for the cases being rolled, pasted, and dried as before, one end may be permanently tied and sealed, or dipped into hot pitch, after which they may be filled in the following manner: first, put in a small quantity of grained powder, which with the rammer and mallet ram down quite hard, then fill up the case as before with the composition, ramming it hard down in the course of the filling, two or three times; this done, cap it with touch-paper, as before directed, and the squib is fit for action.

2. Crackers.—The best material for the cases of crackers is cartridge-paper, the dimensions of which, for an ordinary size, is about fifteen inches long by three and a half inches wide, folded in the following particular manner; it is called *particular,* because on it depends the goodness of the cracker: The method is, first to fold one edge down about three-quarters of an inch broad, then the double edge is turned down about a quarter of an inch, and the single edge is bent back over the double fold, so as to form within, a channel a quarter of an inch wide, which when opened is to be filled with mealed powder, not ground very fine; this powder is then to be covered by the folds on each side, and the whole to be pressed very smooth and close, by passing over it the edge of a flat ruler or some such instrument, and the part containing the powder is to be gradually folded into the remainder of the paper, taking care to press down every fold in the same manner.

The cracker thus far advanced, is to be doubled backwards and forwards in folds about two and a quarter inches, as many times as the length of the paper will allow. After this, the whole should be pressed quite close together by means of a small wooden vice (similar to those known by carpenters under the name of hand-screws, the use of which would be found extremely convenient for many other purposes), and a piece of twine passed twice round the middle across the folds, and the joinings secured by causing the twine to take a turn round the middle at every fold successively. One of the ends of the folds may be doubled short under, which will produce an extra report; the other must project a little beyond the rest for the purpose of priming and capping with the touch-paper; when this is done, the cracker is complete. Crackers, when well made and of sufficient strength, are productive of much mirth, and when of considerable magnitude, furnish excellent means of dispersing a crowd; at the same time they are so perfectly harmless that no evil consequence may be expected to follow the amusement they afford.

3. PIN-WHEELS.—Pin or Catharine wheels are of very simple construction, nothing more being wanted than a long wire former, about three-sixteenths of an inch diameter; on this wire are formed the pipes, which, being filled with composition, are afterwards rolled round a small circle of wood, so as to form an helix or spiral line.

The cases are generally made of Elephant paper, or such as will admit of the greatest length, rolled about four times round the wire, and pasted as they are rolled. When a number of pipes are made and got perfectly dry, they are filled with the composition described at No. 2, in the table. These cases are not rammed, but filled by means of a tin funnel with a long pipe, made so as easily to pass down the case, which is gradually filled by shaking the composition out of the funnel. All the cases prepared being thus filled, one of them being closed at one end, is to be pasted round the flat circle of wood, which must not be above half an inch thick and one inch in diameter, and secured at every half turn by sealingwax. When this is all wound round the circle, and the wheel not sufficiently large, a second case may be inserted into the mouth of the last, taking care that the end introduced is only loosely twisted, otherwise it might obstruct the communication and destroy the effect; but this being properly adjusted, and the joining secured by pasting paper round it, the spiral is to be continued in the same way as before, till the wheel be increased to the proper dimensions, or such as suits the taste of the tyro.

The central block must be pierced in the middle for the purpose of receiving a strong pin, or a small piece of wire, by which the wheel may be attached to a post or any other convenient object; or the pin or wire being inserted into the pith of a hazel-stick, the wheel, without any danger, may be let off in the hand; when the mouth of the last round is primed and capped with touch-paper, on its being lighted, the impulse of the flame against the air forces back the ignited part of the wheel, which continues to revolve till the whole of the composition is consumed.*

* The force by which this wheel revolves is very remarkable, as it unites in itself those two adverse forces which have been the subject of so much mathematical controversy, namely, the centrifugal and the centripetal. It may appear like trifling with science to observe these forces in this simple produc-

4. STARS.—These are small paper globes filled with a composition that emits a most beautiful radiating light, which has been compared to the light of "those endless beauties which adorn our celestial hemisphere." As the purposes for which they are used are chiefly as ornaments to other articles, such as rockets, Roman candles, &c., their dimensions must of consequence be limited or adapted to those articles, therefore their diameters must seldom exceed three quarters of an inch, unless the articles to which they are attached are of more than ordinary dimensions, and for small articles their diameter must be less in proportion. At the beginning of this article, these were called "paper globes," but it must be observed that they are only put in paper when their composition is prepared dry; and instead of paper they are frequently wrapped in a small piece of linen rag, tied closely round with small twine,

---

tion, but that they do exist in it is not less evident; for from the revolutions of the ignited particles of the composition, the former is produced, and from the nature and well-known properties of the evolute curve, *cæteris paribus*, the latter is produced.

The Evolute and Involute Curves are possessed with many remarkable properties, which it would be no difficult task to unfold; but, as it could be of no practical use to the Pyrotechnist, we shall leave it to such of our mathematical readers as are able to appreciate the pleasure which such investigations afford. And for their assistance, we refer them to the excellent writings of Hutton, Simson, Maclaurin, &c.; and to Sect. 4, Book 2, Newton's "Principia," where they will find the subject beautifully illustrated.

and when either of these wrappers are used, a hole must be pierced through its middle, to receive a piece of match left projecting a little on each side.

Though the above mode of making stars is frequently practised, yet it has always been found best to use the composition moist, in the form of stiff paste, when it will not be necessary to inclose the star in anything, because when prepared of such paste it can retain its roundness; there will be no need also of piercing a hole in it for the match, because when newly made, and consequently moist, it may be rolled in pulverized gunpowder, which will adhere to it; this powder when kindled will serve as a match, and inflame the composition of the star, which, in falling, will form itself into stars, and exhibit a most beautiful appearance. For the composition for stars, consult the table Sect. VII., Nos. 3 and 4.

Strung Stars.—To make these, cut some thin paper into pieces of about an inch and a half square; then on each piece lay equal quantities of dry star composition, nearly as much as the paper will contain; then twist up the paper as light as possible. When done, rub some flour-paste on your hands and roll the stars between them; then set them in some warm place to dry. The stars being thus prepared, get some flax or fine tow, and roll a little over each star; then paste the hand and roll them as before, after which set them again to dry; and when that is quite effected, with a spindle make a hole through the middle of each, and thread them on a cotton quickmatch, long enough to contain ten or twelve stars three or four inches distant: by joining sundry lengths of match, any number of stars may be strung together.

TRAILED STARS, or, as they are sometimes called, *cometic-stars*, from their sending out a great number of sparks, which represent a tail similar to that of a comet. There are two kinds bearing the above name, which are those that are rolled and those that are driven; when rolled, they must be moistened with a liquor made of half a pint of spirits of wine, and half a gill of thin size (vellum or any other that is fine), of which use as much as will bring the composition to a consistency proper for rolling into balls; when that is done, sift mealed powder over them and set them to dry.

DRIVEN STARS. — For these, the liquid used for moistening the composition must be spirits of wine, with a little camphor dissolved in it, and but a very small quantity of it; as for driven stars the composition does not require to be wet; cases containing one or two ounces are best for this purpose, which must be made of some very thin paper.

The composition being damped with the spirits of wine and camphor as above, the cases are to be filled and rammed moderately hard, taking care that the case be not broken or the paper driven down on the inside; as a protection to them, while filling and ramming, it will be best to procure several moulds fitted to their exterior diameter. These moulds may be of tin, or any kind of wood, of dimensions suited to stars from eight drachms to four ounces. When they are filled, their cases must be made considerably lighter, which is effected by unrolling the paper within three or four rounds of the charge, which is to be cut off, and the loose edge made fast with a little paste, and afterwards set by for two or three days to dry; when they have attained sufficient dryness they must be cut into lengths proportionate to their weights, which will be nearly as follows: from quarter to half ounce cases, their lengths may be five or six-eighths of an inch; from half to one ounce cases, their lengths may be one inch; if two ounces, one inch and a quarter; from three to four ounces, one inch and a half long: of the smaller pieces, one end must be dipped into melted wax in a manner to cover the composition, the other end must be sprinkled with mealed powder wetted with spirits of wine. Of the larger pieces both ends must be primed with mealed powder wetted as before.

Stars made after the above manner are used almost exclusively for air balloons, and are driven in cases to secure them from the force of the composition with which the balloons are filled; therefore their application to rockets, and other small articles, is quite incompatible with their nature.

Rolled Stars.—These are so called chiefly from the operation employed in their manufacture. Their dimensions are from half an inch to one inch diameter. In the composition care must be taken that the ingredients are well mixed, and previous to its being made up, it must be wetted with the following liquid, sufficient to convert it into paste; spirits of wine one quart, in which dissolve one-fourth of an ounce of isinglass. Too much of the composition must not be prepared at once—one pound will be sufficient for an ordinary number of stars—for if a greater quantity is wetted the spirit will be liable to evaporate, and leave the composition dry and unfit for the purpose, before it can be all rolled up. To make the stars of uniform dimensions, the following method has been found most eligible, and of least trouble: when the composition is properly moistened, roll it with a smooth round stick on any flat even surface, as stone or wood, till its thickness is about half an inch, then divide it accurately into squares, of dimensions suitable to the desired magnitude of the stars: there are other methods for regulating the size of the stars, but this has been found most practicable, and is therefore recommended. Having rolled up the portion of prepared composition as directed, shake them in some mealed powder while they are damp, and set them in some warm place to dry, which will be effected in two or three days; but if wanted immediately, they may be quickly dried in an earthen pan over a slow fire, or in an oven of moderate temperature; when the stars are perfectly prepared they must be preserved in some small box for use, for if exposed to the air they will grow weak, and produce but few of those effects which at other times render them so beautiful.

5. SPARKS.—It is only as it regards magnitude that sparks differ from stars above described, they being generally made of very small size, and consequently of short duration in their exhibition. The method of pre-

paring them is as follows: put into an earthen vessel one ounce of mealed gunpowder, three ounces of powdered saltpetre, and four ounces of camphor, reduced to powder by rubbing it in a mortar with a small quantity of spirits of wine; pour over this mixture some weak gum-water, in which some gum-tragacanth has been dissolved till the composition is brought to a state of thin paste; then take some lint, prepared by boiling it in vinegar or saltpetre, and afterwards dried and unravelled, and put into the composition enough of it to absorb the whole; at the same time take care to stir it well. This matter is to be formed into small balls, of about the size of a pea, which, being dried by a moderate heat, are to be sprinkled with mealed gunpowder, for the purpose that they may readily catch fire.

Another method of making Sparks.—Take some sawdust made from fir, or any kind of wood that burns readily, and boil it in water in which saltpetre has been dissolved; after boiling about a quarter of an hour, the vessel must be removed from the fire, and the liquid poured off, so as to leave the sawdust at the bottom of the vessel, then place the sawdust by itself upon a flat board or table, and while in a moist state sprinkle it with sulphur, sifted through a fine hair sieve: the sifting powder (sulphur) will be improved if there be added to it a small portion of bruised gunpowder. When the whole has been well mixed and of proper consistence, it is to be made up into sparks, as described in the other method.

6. MARROONS.—Marroons are of very easy con-

struction, being nothing more than small cubical boxes, filled with a composition proper for making them burst, and thence producing a loud report, which, and the suddenness of it, is their chief property. They are used principally in combination with other pieces, or to form a battery, in which, by different lengths of quickmatch, they are caused to explode at distinct intervals.

Construction.—Cut a piece of pasteboard into the form represented in Fig. 6, which will fold up into a cubical case, the angles must be well secured by pasting paper over them, the top being left till it is filled; when this is done, the box is to be filled with grained powder, then cement strong paper over the top, and again in various directions over the body; and to increase the strength of the box (which will produce a louder report), wrap round two or three rows of packthread dipped in some strong glue, then make a hole in one of the corners, and introduce into it a piece of quickmatch, and your marroon is ready for action.

Fig. 6.

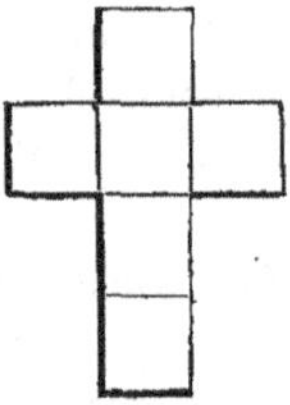

Marroons may be rendered luminous, or caused to emit a brilliant appearance previous to their explosion.

This is effected by covering them with paste made of flour of sulphur, mixed up with thin starch, and afterwards rolling them in pulverized gunpowder, which will serve as a match, or communication; when made after this manner they are called luminous marroons.

7. SAUCISSONS.—These differ only in form from the foregoing articles; till lately no distinction was made between them, nor, indeed, ought any to exist, but the French artists have thought proper to give them the above name from the supposed resemblance they bear to a sausage.

The cases of marroons are made cubical, those for the present articles are made cylindrical, and in proportion must be about four times their exterior diameter in length; their diameters may be from one to two and a half or three inches, and their cases increasing in strength as their dimensions.

The cases must be choked or pinched at one end after the manner of rockets, and tied quite close; and afterwards the former, on which they are rolled, should be pressed hard upon the bottom to make it smooth, and to take out the wrinkles left by the choking; the former, or interior diameter, should not exceed one-half of the exterior diameter of the case.

The cases being thus prepared, they are to be filled with coarse powder one diameter and one-fourth high, and the rest of the paper must be folded down tight upon the powder; then bind them tight in every direction with strong packthread dipped in glue, and leave them to dry as before.

They may be rendered luminous, and the match applied in the same manner as to marroons.

Batteries of Marroons, &c.—These, it has been said, if well managed, will keep time to a march, or a slow piece of music. They must indeed be well managed to do so; individuals have (with care) made several trials, but are seldom fortunate enough to produce such uniformity in their intervals, as to mark correctly their commencement of each bar of music; and if they do not, they fail entirely as to this property. But, however, much effect may be produced by these noisy pieces by arranging them on several stands, with a number of cross rails, on which they are to be nailed, and connected together by means of leaders, &c., of different lengths, according to their distance asunder, observing to use the large and small marroons and saucissons in order to

Fig. 7.

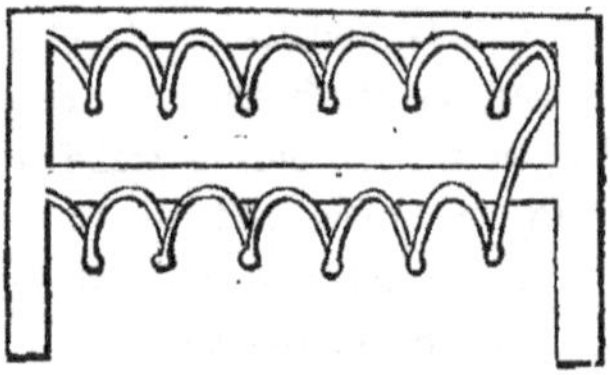

produce a greater variety in the reports, which during the exhibition of other articles is their chief purpose.

A battery with the leaders complete is represented in Fig. 7.

8. Gerbes.—This is a species of firework which, from a cylindrical case, throws up a luminous and sparkling jet of fire, and from its partial resemblance of a water-spout, the French have given it the appellation of gerbe.

Gerbes consist of a strong cylindrical case made of thick paper or pasteboard, and filled with brilliant composition, and sometimes with stars or balls placed at small distances, so that the composition and balls are introduced alternately; immediately below each ball is placed a little grained powder. This last kind of gerbes is more properly called Roman candles, which will be described in the next article. Gerbes are sometimes made wholly cylindrical, and sometimes with a long narrow neck; the reasons for making them with a neck, are deduced from rather philosophical considerations; when fired, they exert great force on all parts of the case, especially at the mouth, proceeding therefrom with great velocity; the reasons therefore deduced for making them with a long neck are—first, that the particles of iron, which enter into their composition, will have more time to be heated, by meeting with greater resistance in getting out, than with a short neck, which would be burnt too wide before the charge be consumed, and spoil the effect; secondly, that with long necks the stars will be thrown to a greater height, and will not fall before they are spent or spread too much; but when they are made to perfection, they will rise and spread in such a manner as to represent pretty exactly the form of a wheat-sheaf.

The diameter of gerbes is generally estimated by the weight of a leaden ball, which the case is capable of receiving; thus they are spoken of as gerbes of eight ounces, one pound, &c. Their length from the bottom to the top of the neck should be about six diameters; the neck being about one-sixth diameter, and three-fourths diameter, long. They are filled in two ways, according as they have a neck, or are wholly cylindrical; the cases of the latter kind are closed below, and are filled like those of serpents, but the composition must be put in by small quantities, and rammed very hard; cases with necks are filled from the bottom, but before commencing the ramming, care must be taken to plug up the aperture of the neck with a piece of wood fitted to its diameter, for if this is not done, the composition will fall into the neck, and leave a vacancy in the case, that will cause it to burst as soon as the fire arrives at that part of it.

It must be observed, too, that the first ramming or two should be of some weaker composition than the body of the case. When filled, the plug must be removed, and the neck filled with some slow charge, and capped with touch-paper; a foot of wood is afterwards to be fixed to the gerbe and well secured, either by a cylinder fixed to the outside of the case, or by having in it a hole, into which the case may be inserted. When either of these methods is employed the foot must be firmly attached.

Sometimes sparks (article 5) are introduced during the filling of the cases, but in this case special care must be taken that they are not broken by hard ramming; their number should be regulated by the size of the case, and when carefully used, they produce a pleasing effect; but they are most adapted to such gerbes as are wholly cylindrical.

The following method of finding the interior diameter of gerbes is generally employed: Supposing the exterior diameter of the case at bottom (which is usually made somewhat larger than the top) to be four inches, then by taking two-fourths for the sides of the case, there will remain two inches for the bore, which will be a tolerable good size, and from the rules given for the height, the same will be about twenty-four inches to the top of the neck. Fig. 8 represents a wooden former; and Fig. 9 a gerbe with its foot complete. The composition for filling will be found in the table, Section VII.

Fig. 8.

Fig. 9.

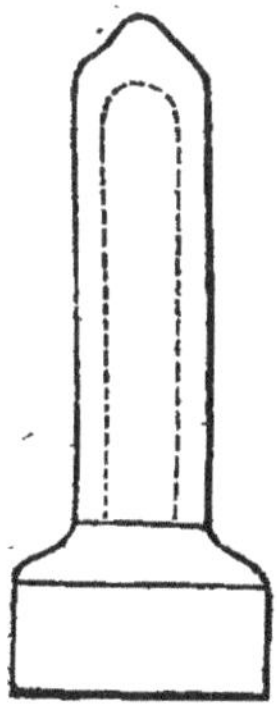

In ramming large gerbes, an external mould will not be requisite, the cases being sufficiently strong to support themselves.

Small Gerbes.—These are frequently called "white fountains;" they differ but little, when used as gerbes, from the foregoing; they are made of four, eight, or twelve-ounce cases, of any length, pasted and made very strong; before they are filled, drive in, about one diameter of their orifice high, some good stiff clay, and when the case is filled, bore through the centre of the clay to the composition, a vent-hole of common proportion, which must be primed and capped as before.

These cases are sometimes filled with Chinese fire; in this case clay must not be used, but filled the same as cylindrical cases, and footed and primed in the same manner.

9. ROMAN CANDLES.—Roman candles are constructed nearly after the manner of gerbes; their cases are made perfectly cylindrical, as above described, and between the layers of composition are interposed balls, or stars, which are prepared as directed in article 4. In filling and ramming Roman candles, especial care must be taken that the stars are not broken in the operation. When the cases have been properly rolled and dried, and their bottoms firmly secured by tying them with some strong twine, it is best, previous to putting in the composition, to ram a little dry clay, which will fill up the hollow, and leave a better bottom to the case. This being properly done, put in a small quantity of corn powder, and over this a small piece of paper, just to prevent the composition from mixing with the powder; then as much of the composition is to be put in as will, when rammed hard down, fill the case about one-sixth of its height; then over this a small piece of paper (covering about two-thirds of the diameter) as before, then a little corn powder, and upon that a ball is to be placed, observing to let the ball be somewhat less than the diameter of the case. Over this first ball more of the composition is to be introduced, and pressed lightly

down till the case is about one-third full, when it may be rammed, but with gentle strokes, lest the ball be broken by it; then a piece of paper, a little corn powder, and upon it another ball, as before; so that the case after this manner will contain five or six balls with regular beds of composition between them, and have about the same length of composition above the highest ball. When the case is thus filled it is to be capped with touch-paper by pasting it round the orifice; and a little priming of mealed powder being added, the piece is rendered complete.

In regard to the stars or balls, it is best that their form be flat and circular, or even square rather than spherical, as they will be less liable to be injured in the filling; they should also be somewhat different in size, which is found to add much to their effect; that is, let the first star be about two-thirds the diameter of the case, let the next be a little larger, and so on increasing to the fourth, fifth, or sixth, which last should fit tight into the case.

Observe also to let the quantity of powder at the bottom of each ball increase as the balls increase in diameter, or as they come nearer the top of the case; not on account of the additional weight of the ball, but, as on those balls situate near the top, the force of the powder ceases to act on the ball, sooner than on those situate lower in the case, consequently the force to throw the ball to the same distance must increase proportionably; another reason for decreasing the quantity of powder towards the bottom is, that the same quantity used with the bottom as with the top ball, would cause the cases to burst, and destroy all the effect which they are intended to produce.

The composition for filling will be found in the table, Section VII.

The best way of exhibiting these Roman candles is to place them in rows on a stand, some fixed quite perpendicular, others declining at different angles, that the balls may be projected to various distances, and produce a more beautiful effect. The greatest angle of declination should not exceed forty-five or fifty degrees.

A very pleasing variety of gerbes may be produced by filling the cylindrical cases with the composition called Chinese fire (see next article): being filled with red or white, and used with different proportions of the ingredients, they may be cast into many and various shades of colors.

10. Chinese Fire.—The principal ingredient which forms this beautiful composition, has been already described in Section II. under the name of iron-sand; what is given in this place is, the proportion in which it is used with the other ingredients; the composition is rendered into two particular distinctions, namely, red, and white, and each of them made with different proportions of the ingredients according to the calibers of the cases intended to be filled with it, which caliber is estimated by the weight of lead balls which will just fill out their diameter, as was taught in the article Gerbes.

For Red Chinese Fire.

	Calibers.	Saltpetre.	Sulphur.	Charcoal.	Sand 1st order.
I.	12 to 16 *lb.*	1 *lb.*	3 *ozs.*	4 *ozs.*	7 *ozs.*
II.	16 to 22 *lb.*	1 *lb.*	3 *ozs.*	5 *ozs.*	7 *ozs.* 8 *drs.*
III.	22 to 36 *lb.*	1 *lb.*	4 *ozs.*	6 *ozs.*	8 *ozs.*

For White Chinese Fire.

	Calibers.	Saltpetre.	Bruised powder.	Charcoal.	Sand 3d order.
I.	12 to 16 *lb.*	1 *lb.*	12 *ozs.*	7 *ozs.* 8 *drs.*	11 *ozs.*
II.	16 to 22 *lb.*	1 *lb.*	11 *ozs.*	8 *ozs.*	11 *ozs.* 8 *drs.*
III.	22 to 36 *lb.*	1 *lb.*	11 *ozs.*	8 *ozs.* 8 *drs.*	12 *ozs.*

After carefully weighing the several ingredients, observe to sift the saltpetre and charcoal two or three times through a hair-sieve in order that they may be well mixed; the iron-sand is then to be moistened a little with brandy or spirits of wine, which will make the sulphur adhere, and they must be well incorporated. The sand, now said to be sulphured, must be put to the mixture of saltpetre and charcoal, and then stirred and turned about till the parts are thoroughly incorporated.

SECTION VI.

ROCKETS.

THIS part of the work treats on the most beautiful of all pyrotechnic productions.

Rockets have ever held the first place among single fireworks since the invention of the art; and to which they are justly entitled, both for the pleasing appearance they produce when fired by themselves, and the extensive application of them to increase the beauty of the other exhibitions.

They are called by the Italians, *Rochette* and *Raggi;* by the Germans, *Ragetten* and *Drachetten;* by the French, *Fuzees;* and by the Latins, *Rochetæ;* from which appears to be derived the name given them by the English. So much for their names: as to their invention it is most probable that it took place at a very early period, if not among the first productions of the art. By the ancient pyrotechnicians, they were considered as the most difficult articles of manufacture, insomuch that it was the first task enjoined to the disciples of Prometheus,* or professors of the art; and the

* He is fabled by the ancients to have formed a man of clay, or earth, and to have stolen fire from heaven, with which he animated the man he had made.—*Heathen Deities.*

goodness of the article furnished a criterion of their pretensions.

It is to be questioned whether the ancients had such a variety of these articles as there now are; but it is pretty certain that they were well acquainted with the proper proportions of the moulds requisite for their manufacture, insomuch that in many of their treatises, they employed the most difficult mathematical calculations, and gave intricate algebraic formulæ, for the purpose of finding their true proportions; but many of such useless difficulties may be avoided, and the explications rendered familiar, without wholly sacrificing scientific investigations.

Rockets consist of strong paper cylinders, which being filled with the proper composition rammed hard, and fire being applied to their apertures, they are caused to ascend into the air, or in any required direction: they have generally a head fixed to them, containing corn powder, sparks, and many other decorations, which, when the body of the rocket is consumed, take fire, burst in the air, and produce a most beautiful appearance: these are called sky-rockets. Others are made to run with a great velocity along a line, and are therefore called line-rockets, or courantines. Some are fixed on the circumference or on the axle of a wheel, and are denominated wheel-rockets; while another kind have their cases made perfectly water-tight, and being filled with stronger composition, admit of being plunged in and under water without retarding their inflammation; these receive the significant appellation of water-rockets.

1. *Sky-rockets.*—Sky-rockets, in regard to size, are divided into three kinds, namely, those the caliber or internal diameter of which does not exceed that of a pound bullet; or having their orifice equal to a leaden bullet, which weighs exactly one pound; for the relative magnitude of rockets is estimated by the diameter of lead balls or bullets, after the manner taught in the article gerbes. Those, therefore, the caliber of which does not exceed a pound bullet, are termed small-size rockets; those whose caliber is from one to three pounds are of the middle size; and those whose calibers exceed the last dimensions, are termed rockets of the largest size; or are named after their weight, estimated as above.

The moulds and apparatus requisite for making rockets shall now be described, for on the due proportion of which (as before observed) depends much of the goodness of the article. These moulds are also requisite in order that any number of rockets may be prepared of the same size and force. As rockets are made of various sizes, it is evident that moulds of different diameters must be produced.

Fig. 10 represents a mould made and proportioned by the diameter of its caliber, which is divided into equal parts and rendered into scale, by which the relative proportions may be understood, merely by a contemplation of the figure. Thus A B is the caliber, or diameter; C D its whole height, including the foot complete, and equal to eight diameters, as per scale: E is the thickness of the mould, and may be about half a diameter; it should be made of some hard wood, such as lignum vitæ, or box, and may be either ornamented or plain; F is an iron pin, which serves to fix the cylinder firm to its foot. Fig. 11 is the foot detached from the cylinder, and drawn in true proportion, as per scale; G, H, I, J, is the base, and may be about one

Fig. 10.

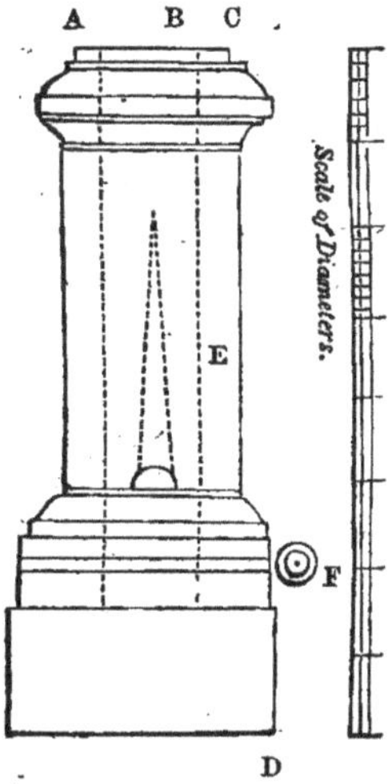

Fig. 11.

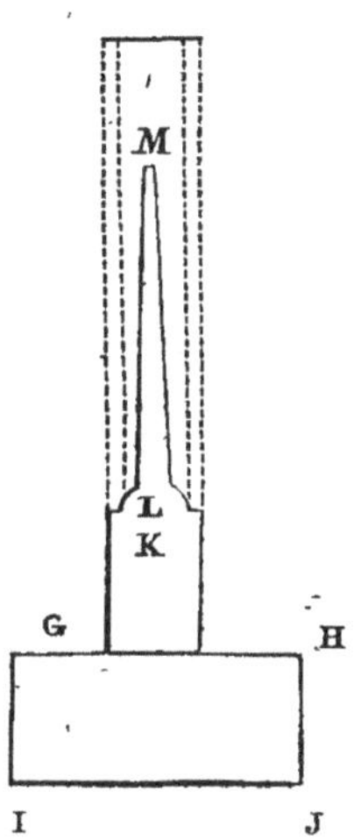

and a half diameter high; K, the choke, which serves to connect the cylinder to the foot; L is the nipple, which is half a diameter high, and in thickness equal to the former, or five-eighths diameter; M is the spindle, whose height is three diameters and a half from the nipple, and at the bottom one-third or fourth diameter, from thence tapering to one-sixth diameter in thickness:

This spindle should be of gun-metal, and inserted firmly into the bottom; its purpose is to preserve a vacuity in the centre of the charge, the nature of which will be hereafter explained. Fig. 12 is a former in two pieces, connected by an iron pin (in diameter equal to the bottom of the spindle), to which both ends are rounded off, in order that the choke or contraction in the case may be effected more easily; the diameter of this former must be the same as that of the nipple, or suppose the diameter of the mould be divided into eight equal parts (which is done on one part of the scale), then the diameter of the former must be equal to five of these parts.

Fig. 12.

The length of this former, or roller, is not particular, provided it be long enough to admit of good hand-hold in the rolling of the cases; the short part of the former, A, may be two diameters in length, and should have a line, B, marked round it in the middle, or one diameter from the end; the longer part may be seven or eight diameters, which will give good hand-hold in the rolling.

Figs. 13 and 14 are drifts used in loading the cases, which must be bored lengthwise to fit on to the spindle.

Fig. 13. The first drift should be bored the whole length of the spindle, the second should be bored one and a half diameter; when the case is loaded and rammed above the spindle, a short solid drift must be used, and these drifts should be a little less than the former, to prevent injury to the inside of the cartridge, when

Fig. 13.

Fig. 14.

driving in the charge. They should be made of some hard wood, and their extremities secured by ferrels of brass, or any other metal, which will keep them from splitting or extending: their lengths are of little consequence, provided they do not much exceed the relative depths of the case; for, as the workmen say, the longer the drift the less will be the pressure on the composition by the blow given by the mallet.

The proportion between the length of rockets and their caliber, is not the same in rockets of greater or less dimensions than those given above, but should vary nearly as their magnitude; that is to say, their length should be diminished as their caliber is increased. The length of the mould for small rockets should be six times the caliber, but for rockets of the mean and larger size, it will be sufficient if the length of the mould be five times, or even four times that of the caliber.

The following is a table computed to regulate the height and diameter of the mould according to the weight of the rockets, when they are driven solid, or without the use of a spindle. It is extracted from an old treatise on fireworks by Lieutenant Robert Jones, and inserted for the assistance of those who may wish to construct rockets without the spindle, a practice not recommended to those for whom this work is designed. To those who manufacture fireworks for sale, it is certainly the most expeditious method to ram them solid, and with the machine to bore or pierce them afterwards; but to those who make rockets for their own private recreation, it is by far the most eligible to load them over a spindle, for by the other method it will require a very expensive apparatus,* and at first more skill to use it than the tyro will possess, and at last he will never be certain that he has made a good article.

---

* Fig. 15 represents an apparatus for boring rockets when driven solid. Should this method be attempted, it is one of the simplest kind that can be used. A, B, are two movable blocks hollowed out at their edges to receive the rocket. C, D, are two screws in each, going through the sides of the frame

TABLE I.

Dimensions of Rockets.

Weight of rockets.	Length of the moulds without their feet.	Interior diameter of the moulds.	Heights of the nipples.
6 *lbs.*	34.7 *inches.*	3.5 *inches.*	1.5 *inches.*
4 do.	31.6 do.	2.9 do.	1.4 do.
2 do.	13.3 do.	2.1 do.	1.0 do.
1 do.	12.2 do.	1.7 do.	0.85 do.
8 *ozs.*	10.12 do.	1.3 do.	0.6 do.
4 do.	7.75 do.	1.12 do.	0.5 do.
2 do.	6.2 do.	0.9 do.	0.45 do.
1 do.	4.9 do.	0.7 do.	0.33 do.
½ do.	3.7 do.	0.55 do.	0.25 do.
6 *drs.*	3.5 do.	0.5 do.	0.22 do.
4 do.	2.2 do.	0.3 do.	0.2 do.

Fig. 15.

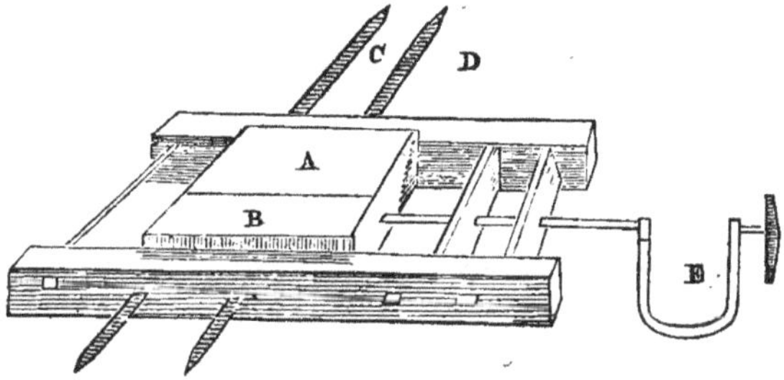

to make them fast. E, is a brace carrying a shell-bit, in dimensions proper for the rocket. Fig. 16, is a similar bit fixed in a handle; which is useful in cleaning out the bore when the rocket is taken from the blocks.

Fig. 16.

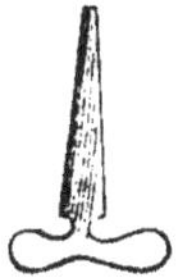

By this table it will be seen that a six-pound rocket rammed solid, must be thirty-four inches seven-tenths in length; its external diameter three inches five-tenths, or three and a half inches, and the height of the nipple one inch and a half. The diameter of the nipple in this and all other cases, must be equal to that of the former, and in regard to its height it answers best when the cavity which it formed at the mouth of the rocket is hemispherical, or equal in height to half its diameter.

The following table shows the method of finding the caliber of rockets according to their weight, which is computed by the principles already given; that is, a pound rocket is such that its aperture will just admit a bullet of a pound weight, and so of the rest.

TABLE II.

Of the Caliber of Rockets of a Pound Weight and below.

16 *ounces.*	19½ *lines.**	14 *drams.*	7¼ *lines.*
12 do.	17 do.	12 do.	7 do.
8 do.	15 do.	10 do.	6⅓ do.
7 do.	14¾ do.	8 do.	6¼ do.
6 do.	14¼ do.	6 do.	5⅔ do.
5 do.	13 do.	4 do.	4½ do.
4 do.	12⅓ do.	2 do.	3¾ do.
3 do.	11½ do.		
2 do.	9⅙ do.		
1 do.	6½ do.		

* A line is the twelfth part of an inch, or 144th part of a foot. Geometricians conceive the line, notwithstanding its

The use of *this* table will be easily understood; for, as in the first instance, if a rocket of sixteen ounces ought to be nineteen and a half lines in diameter, one of twelve must be seventeen lines, one of eight ounces fifteen lines, one of eight drams six and a quarter lines; and so of the others.

If the diameter of the rocket be given, it will be as easy, by the reverse method, to find the weight of the ball corresponding to that caliber. For example, if the diameter be fifteen lines, it will be immediately seen, by seeking for that number in the column of lines, that it answers to a ball of eight ounces.

As the foregoing table extends only to rockets of sixteen ounces, or one pound, and from that downwards, the following will be found equally useful for those of superior dimensions.

---

smallness, to be subdivided into six points. The numbers in the table might have been given in lines and points, but it was thought the fractional numbers would be as well understood.

TABLE III.

Of the Caliber of Moulds from 1 *to* 57 *Pounds Ball.*

Pounds.	Caliber.	Pounds.	Caliber.	Pounds.	Caliber.
1	100	20	271	39	339
2	126	21	275	40	341
3	144	22	282	41	344
4	158	23	284	42	347
5	171	24	288	43	350
6	181	25	292	44	353
7	191	26	296	45	355
8	200	27	300	46	358
9	208	28	304	47	361
10	215	29	307	48	363
11	222	30	310	49	366
12	228	31	314	50	368
13	235	32	317	51	371
14	241	33	320	52	373
15	247	34	323	53	376
16	252	35	326	54	378
17	257	36	330	55	380
18	262	37	333	56	382
19	267	38	336	57	385

By this table, the weight of the ball being given, the size of the mould may be found after the following manner. Suppose it to be 18 pounds; opposite to it, in the column of calibers, is 262; then say by the rule of proportion, as 100 is to 19½, so is 262 to a fourth term, viz., 51.09, which is the number of lines of the caliber required: therefore, the caliber of a rocket of eighteen pounds, will be fifty-two lines nearly, or four inches and four lines. The caliber may be also found by multiplying the number answering to the pounds by 19½, and cutting off from the product the two last figures; thus suppose the number be 252, which multiply by 19½, the product 4914 separated by the decimal point will give 49.14, or four inches a line and one-eighth.

Now suppose the caliber be given in lines, the weight of the ball may be found with equal ease, *e. g.* if the caliber given be 36 lines, then as 19½ : 100 : : 36 : 184; the nearest number in the table to this is 181, which shows that the weight of the ball will be rather more than six pounds; therefore a rocket, the caliber of which is thirty-six lines, is a rocket of a six pound ball.

Remarks on the foregoing Tables.

TABLE I. gives the dimensions of rocket-moulds when the rockets are rammed solid; it was calculated, as its author states, from repeated experiments; it is inserted for the information of the reader, but the method of solid ramming is not recommended for practice.

TABLE II.— This table may be perfectly understood by the explanation given of its use, and from considering that a lead bullet of a pound weight, is just nineteen and a half lines in diameter, as may be proved by experiment; the inferior numbers are likewise the diameters of the inferior weights.

TABLE III.—This table is only an extension of the latter, although its arrangement is somewhat different; for if nineteen and a half, the diameter of a ball weighing one pound, be assumed as unity with any number of ciphers, answering to the number of parts which the same diameter is divided into (which may be done by means of the diagonal scale), let this number be 100, which answers to one in the column of pounds: that is to say, if 100 be assumed for the first number, and it be raised to a third power,* the first cube will be 1,000,000, the cube root of which (being 100) must be placed in the table as the first root, and answering to unity in the column of pounds: then for the second number, which is two pounds, the cube root of double that number must be extracted, viz. 2,000,000, which will be 126 nearly (or continued to more places 1,259,-921), and this will be the second number in the table; and in the same manner will the third number be found, that is, by trebling the first cube and extracting the root as before, which will be 144, and so of the fifth, sixth, &c., to the end of the table. These tables are indispensable in the making of rockets, in order to preserve uniformity in rockets of the same kind, and to render more certain their effects, as has been corroborated by repeated experiments.

Preparing the Cases.—For this purpose large stiff paper of a particular kind is to be used; namely, that which is known by the name of *cartridge-paper*. For cases from the smallest size, up to four pounds, this is the best material that can be employed; it must be

* Euclid 12, 18. Spheres are to each other as the cubes of their diameter, this is the principle employed in the construction of the table; but the method is converse, being that of extracting the roots.

wrapped round the former (whose proportion to the mould has already been given), till it fits tight into the cylinder, and the last fold secured by means of common paste. If some thin paste is used throughout the rolling, the cases will be much improved.

For rockets of a larger size, the cases must be made of some stronger material, such as pasteboard, of the thin and inferior kind, the folds of which must be well secured with some strong paste or glue. In making the cases a pattern of the outer fold, with one end sloped off, should be preserved for each size, and on it marked the number of sheets or folds requisite to make that size case. This method will help to insure a regularity in the make and formation of the cases.

The general Method of Case-making.—The paper being cut to the necessary size for the case required, lay it on a table, and with a brass former take one turn very tight round the paper, paste it thinly all over, and roll it up with what is termed a rolling board (a piece of deal about two feet long and nine inches wide with a handle in the centre), until it is of the required size. It is now to receive the contraction, or as it is generally termed, the choke; which is effected by the simple apparatus represented by Fig. 17. Let the former and small end-piece be now joined by means of their connecting wire, and let the short piece be thrust into the case as far as the line B, marked round it for the purpose; then pass the cord once round the case, exactly over the juncture of the formers, and at first press gently with the foot on the treadle, and keep rolling the case on the line, which will cause the choke to be free from wrinkles and other inequalities. A small piece of wood (termed a nipple) is placed in the end of the for-

Fig. 17.

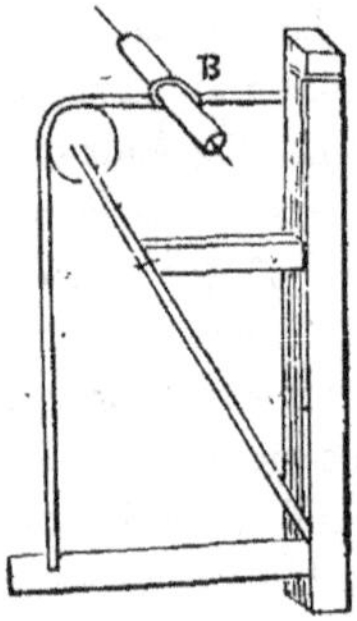

mer, of the size required, which for rocket-cases is *one-third the interior diameter*, and *for fixed cases one-sixth*.

Cases of small dimensions may be easily contracted after the above manner, but when of larger size, they will present more resistance to the choking cord than it will be able to overcome: this difficulty may be obviated, by moistening with water the end of the case, and choking it previous to the envelop of the last sheet; which may then be put on, and again choked, and the contraction well secured by twine, or strong waxed thread, which must be passed several times round the case, and afterwards secured by two or three running knots made one above another.

The case (still remaining on the former), is now to

be inserted into the cylindrical mould without its foot, and set upon some solid block, and the former driven hard upon its end-piece, so as to make the contraction smooth and close; after which the case is to be cut to its proper length, so as to rise a little above the mould, and allowing half a diameter from the choke to the edge of the mouth. The cutting the case to its proper length will be best effected while on the former, which, when done, the former is to be pulled out, and the case being put again into the mould, having the foot and spindle properly fixed to it, must be driven down upon the spindle with the long perforated drifts, so as to make the contraction of the proper size.

Filling and ramming the Cases.—In this part of the operation, as much care must be taken as in any of the past; for if the least inequality exist in the density of the composition, produced by inattention to the ramming, the rockets will not rise with an uniform motion, nor ascend to their proper height; but on the contrary, will observe a very erratic motion, and be deflected by every renitent particle they may meet with in their course.

To avoid this disappointment, and to render more certain the ascent of the rockets, the following directions must be attended to:—

1. The composition must not be too dry, or it will be liable to disperse, and fly about in a kind of subtile meal, or dust, while driving it; but if moistened a little, just to destroy its dusty nature, with some of the liquid mentioned in the early part of this work, it will cause

it to collect, and be more solidly compressed in the case of the rocket.

2. No more of the composition should be put into the case at each ramming, than will cause it to rise one-half of its interior diameter; and the filling must be thus gradually continued, till the charge rises exactly one diameter above the spindle.

3. Much has been said by writers on Pyrotechny respecting the number of blows proper to be given to the drift, to each ladleful of composition (a piece of copper made into the form of a scoop, and holding the proper quantity answers best for a ladle); some have assigned to rockets of four ounces sixteen strokes with the mallet, to those of one pound twenty-eight strokes, and so increasing the number of strokes by six, to every pound; but these rules are more minute than useful; for the same mallet, by possessing a different momentum, might produce an effect, at one time *double*, *treble*, or perhaps *less*, than at another. It is therefore impossible to assign any determinate number of strokes, to be given at each ramming; the only certain rule is, that the composition ought to be driven till it becomes quite firm and compact, and that its density (as near as possible) be the same throughout the whole of the charge. If it be thought that the rules for the number of strokes assist in any way to impart this property to the charge, they can of course be resorted to.

4. In ramming, it is best to keep the drift constantly turning round in the case; and in using the perforated drift, be sure to knock out the composition from the

hollow, every ramming, or it will be liable to be split by the spindle.

5. Invert the case at the close of each ramming, in order that the loose particles of the composition which are not compressed may escape, for if suffered to remain in they would prove injurious.

6. Rockets should always be rammed on a solid block, or on a post set fast into the earth; their ramming cannot be properly effected on any table whatever.

7. Rockets must be rammed with mallets somewhat proportionate to their magnitude; that is, if a rocket of one pound can be properly rammed with a mallet weighing two pounds, a rocket of two pounds should be rammed with a mallet of four pounds, or nearly in that proportion. Rockets above eight pounds cannot well be rammed by hand; but when wanted of such magnitude they must be rammed by means of a machine similar to that used for driving piles into the earth, and familiarly called *a monkey*. Rockets of large dimensions, whose cases are made of strong material, properly prepared, may be conveniently rammed without being placed in a cylinder, which will be an advantage, as so many moulds will not be required. But for this method of ramming some *brass* or *iron* nipples, of the size proportionate to the rocket should be used, which should be made to screw into one part of the driving block; and for the purpose of making the case more firm upon it while ramming, a stake or upright piece must be made firm to the block, standing up the height of the case, and at a suitable distance from the nipple: the side of

this stake next the case must be fluted out so that the case will fit closely into it. On the opposite side of the case must be applied a loose piece, fluted in a similar manner; then with a cord tie the case and two half moulds (which these two pieces will nearly form) together, and the case will be ready for filling. The cases being filled to the proper height, *i. e.*, one diameter above the spindle, if the rocket is to be without furniture, separate with a wire of any kind, half the folds of the paper which remain above, and having turned them back on the composition, press them down with the rod and mallet in order to make them smooth and even. Then pierce three or four holes in the folded paper by means of a spindle, which must be made to penetrate to the composition of the rocket. These holes are for the purpose of forming a communication between the body of the rocket and the vacuity at the extremity of the carriage, as it is called, or that part which has been left empty. In small rockets this vacuity is filled with granulated powder (which serves to let them off when their charge is consumed); they are then covered with paper, and either pinched quite close by means of the choking apparatus, or crowned with a little conical cap, which will cause them to ascend to a greater height. If one hole only is made in the centre of the folded paper, it will answer the purpose of three or four, taking care that it be as straight as possible, and about one-fourth the diameter of the caliber of the case; in this hole a little of the composition of the rocket should be put, that the fire may not fail to be communicated: a rocket finished after this manner is represented in Fig. 18. In rockets of larger dimensions instead of granulated powder, the coffins, or pot containing the stars, serpents, petards, &c. are adapted to the top of the case. The petard is a small round box of tin-plate united to the diameter of the case, and filled with fine gunpowder; it is deposited on the composition after the ramming, and the remaining paper folded down over it to keep it secure; the petard produces its effect when the rocket is in the air and the composition is consumed. The other furniture is attached to the rocket by adjusting to its head an empty pot or case of larger dimensions than itself, in order that it may contain the various appendages, which are to render it so superior to the others, in the beauty and splendor of its emication.

Fig. 18.

Preparing and fixing the pots to the heads of rockets.—Rockets which have furniture attached to them, are rammed somewhat different to those which are without any appendages, but the difference is only in this particular; when rammed one diameter above the spindle, instead of turning down the inner folds of the paper upon the composition, ram on to the composition one-third diameter of pure dry clay, and through the centre of it bore a hole (about one-fourth diameter) and put into it a little of the composition, in order that the charge may communicate with the powder, &c., in the head.

The head of a rocket must be about two diameters high, and one diameter one-sixth wide. The case must be rolled upon a former, having, at the end opposite the handle, a square indent, corresponding in thickness and width of the collar, as is represented in Fig. 19. Fig. 20 is the collar, turned out of lime-tree, poplar, or any

Fig. 19.

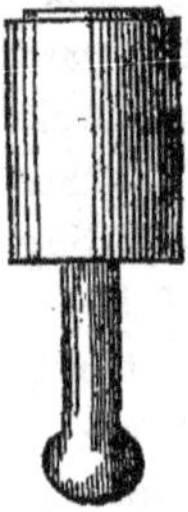

Fig. 20.

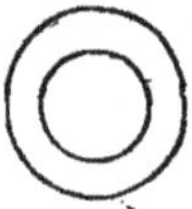

light wood; its exterior diameter must be equal to the interior diameter of the case, or the same as the former, and its interior diameter, not quite so wide as the interior diameter of the rocket case; in thickness it should be equal to one-sixth diameter, and round its edge should be a groove, so that the case for the head may be firmly fixed to it. To form the case, three or four rounds of paper or pasteboard must be rolled round the former with the collar on, and well secured by paste; the end over the collar is to be pinched, by means of the cord and choking apparatus, into the groove in its edge, and afterwards secured by some twine tied closely around it. The purpose of the collar is to keep the head in a proper form, to make a bottom for the filling of it, and to make it more firm and better connected to the case. When the head is thus made, being properly fixed to its collar, it is to be made fast, by means of ordinary glue to the top end of the rocket, in which operation the reason and use of making the interior diameter of the collar less than the exterior of the case will plainly appear; it will be evident that the case of the rocket will be too large for the former, without some alteration, which alteration must be made in the following manner: Mark round the diameter of the rocket the proper distance from the top, or so that the collar is about its thickness above the ramming of the case, and take off about three rounds of paper, which will leave a shoulder to the case, on which the collar may rest, and be made quite secure by pasting paper round their joinings beneath.

The manner of charging the pot above described must be left to the tyro himself, it depending chiefly on his taste and wishes, as he may either fill it with serpents, crackers, saucissons, marroons, stars, sparks, showers of fire, or anything to which its capacity is adapted; it will be best, however, to unite several of the different articles in one head, that the beauty of the exhibition may be increased.

In the filling of the head, the following directions must be observed:—

The paper over the charge of the rocket must be pierced, and a little of the same composition shook into the holes; then arrange in the head the different articles with which it is to be charged, but take particular care that the quantity introduced is not heavier than the body of the rocket. When the head is loaded, a few balls of paper should be put round the different articles, so as to keep them properly in their places. At the top part of each head put a ladleful of mealed powder (the ladle used in filling the cases is meant), which will be enough to burst the head and disperse the stars or whatever it contains.

In loading the head with cases of any kind, be sure to place their mouths downwards, without any touch-paper; the head may be nearly filled with the articles they are loaded with, after which paste on the top of them a piece of ordinary paper; and over this must be placed a cone of the same material, made upon the conical former, Fig. 21. To make the caps, describe

Fig. 21.

(with a pair of compasses opened to the length of the former) a circle, which being divided into two equal parts, will make two caps; over which must be pasted another similar cap, but of larger dimensions, to extend below the bottom of the inner one; so that, being just clipped a little, and applied to the head, it may be pasted to it, which will be a sufficient fastening.

The last business in the manufacturing of a rocket is that of fixing it to its rod, which shall now be described;

as much nicety being required in it as in any of the past operations.

The rod should be made of a clean piece of fir, perfectly straight, and its dimensions regulated by the size of the rocket, in such manner that, when suspended on the edge of a knife or wire, about an inch from the choke, the rod and rocket shall be in equilibrium. The following table has been computed for the lengths and proportions of the rod, and may be relied on :—

Weight of the Rockets.		* Length of the Rods.		Thickness and Width at top.			Square at Bottom.	
lb.	*oz.*	*Feet.*	*Inches.*	*Inches.*			*Inches.*	
6	0	14	2	1½	by	1⅞	0	¾
5	0	13	8	1¼	—	1¾	0	⅜
4	0	12	9	1¼	—	1½	0	⅝
3	0	10	8	1 1/7	—	1⅛	0	½
2	0	9	3	1⅛	—	1	0	½
1	0	7	10	¾	—	⅞	0	⅜
0	8	6	6	½	—	¾	0	¼
0	4	5	2	⅜	—	⅝	0	¼
0	2	4	1	3/10	—	½	0	3/16
0	1	3	5	¼	—	⅜	0	3/16
0	½	2	3	3/16	—	¼	0	⅛
0	¼	1	10	⅛	—	3/16	0	⅛

By the above table, it will be found that a rocket of six pounds will require a rod fourteen feet two inches

* When the rods are of large dimensions, they should be bored at the top, and filled with powder, which blows them to pieces previous to their return to the earth, and prevents any mischief which might happen by their falling otherwise.

long, which being properly planed to the other dimensions, it is to be hollowed out on the side next the rocket; and on the side opposite, two notches must be made, one about an inch from the end (the rod going up to the under side of the head), and the other opposite the choke of the rocket, in order to admit the string with which it is tied, and that it may be more firmly attached to the rod. Although the foregoing table has been carefully computed, and that from experiment, yet it will not be well to depend entirely upon it, but rather to produce an equilibrum between the rod and rocket (by means of a lighter or heavier rod), when suspended as before. It is of consequence that this be attended to; for without a proper equilibrium, the rocket will ascend in an oblique direction, and fall to the ground long before its composition is consumed.

Fig. 22.

In firing these rockets, two fixed rings must be screwed fast into an upright post, and exactly opposite to each other, the upper one near the top of the post, and the other about two-thirds the length of the rod downwards; the rod must be passed down them, and the mouth resting lightly on the upper one: the rocket must be quite free from the post. When thus fixed, and a lighted port-fire applied to its mouth, it will (if properly made), immediately ascend with a prodigious velocity, and having attained its greatest height will there burst and discharge its luminous beauties in the atmosphere. A rocket with its head and rod complete is represented at Fig. 22.

For mixing the Composition.

The charcoal, which in commerce is purchased in the stick (that made of alder, willow, or dogwood being the best), is to be reduced to a state of powder, by rolling an iron or brass weight over it until reduced to a sufficient state to pass through a wire sieve of sixteen meshes to an inch, the saltpetre being reduced by heat to a state of pulverization, and the sulphur by sublimation, a process here unnecessary to explain, those ingredients being sold at most chemists ready for use.

The saltpetre and sulphur are first separately passed through a fine hair-sieve, and then well mixed together with the hands; afterwards passed three times through the same sieve; the charcoal is then to be thinly spread over the mixing-table and the other ingredients sifted over it, and then passed altogether through the wire-sieve.

Proportions of Composition.

1. For one and two ounce rockets, the ingredients for a proper composition should be: One pound of gunpowder, two ounces of soft charcoal, and one ounce and a half of saltpetre.

2. Two to three ounce rockets: To four ounces of gunpowder add one ounce of charcoal, or to nine ounces of powder add two ounces of saltpetre.

3. Four ounce rockets: To one pound of gunpowder

add four ounces of saltpetre and one ounce of charcoal. The composition will be much stronger if in this proportion: To ten ounces of powder add three ounces and a half of saltpetre and three ounces of charcoal.

4. Five or six ounce rockets: Gunpowder two pounds five ounces, saltpetre half a pound, sulphur two ounces, charcoal six ounces, and iron filings two ounces.

5. Seven or eight ounce rockets: Gunpowder seventeen ounces, saltpetre four ounces, sulphur three ounces.

6. Eight to ten ounce rockets: Gunpowder two pounds five ounces, saltpetre eight ounces, sulphur two ounces, charcoal seven ounces, iron filings three ounces.

7. Ten or twelve ounce rockets: Gunpowder one pound one ounce, saltpetre four ounces, sulphur three and a half ounces, charcoal one ounce.

8. Twelve to fourteen ounce rockets: Gunpowder two pounds four ounces, saltpetre nine ounces, sulphur three ounces, charcoal five ounces, iron filings three ounces.

9. One pound rockets: Gunpowder one pound, charcoal three ounces, sulphur one ounce.

10. Two pound rockets: Gunpowder one pound four ounces, saltpetre two ounces, charcoal three ounces, sulphur one ounce, iron filings two ounces.

11. Three pound rockets: Gunpowder four ounces, saltpetre one pound, sulphur eight and a half ounces, charcoal two ounces.

12. Four pound rockets: Gunpowder half a pound, saltpetre fifteen pounds, sulphur two pounds, charcoal six pounds.

For rockets of the largest size: To eight pounds of saltpetre add twenty ounces of sulphur and forty-four ounces of charcoal.

The ends of rockets (next the head) have a ladleful of *clay* driven on the top of the composition; it is prepared by drying and grinding, taking great care that it is perfectly free from any grit.

It will be well to describe some of the various modifications which rockets are susceptible of in their exhibitions; in which it will be endeavored to blend the most prominent features: to attempt to give the whole, would annihilate the purpose of the work; indeed, it is impossible to set limits to the field of variety which here opens itself; a few of the most particular only will therefore be described, and the rest left to the tyro, which will afford him a pleasing source of amusement, and furnish excellent matter for the exercise of his ingenuity.

1. *To cause a Rocket to ascend in a Spiral Form.*

The rod of a rocket has been compared to the rudder of a ship or the tail of a bird; the purpose of which is to make the ship or bird turn towards that side to which it is inclined; a straight rod, as experience proves, causes a rocket to ascend in a straight line, because the centre of gravity lays in, or is parallel to, the centre line of the rod; but if a crooked rod be applied, or one that forms part of a circle, this will not be the case, for the first effect will be to make the rocket incline towards that side to which it is bent; but the centre of gravity bringing it afterwards into a vertical position, the result will be that the rocket will ascend in a spiral form.

Rockets exhibited in this way, evidently displace a greater volume of air, therefore, as a consequence, they cannot ascend so high as those that are impelled in a straight direction; but, nevertheless, their peculiar flight will produce a very pleasing effect.

2. *Towering Rockets.*

So called from their ascending to a greater height than any others. This is effected by fixing one rocket on the top of another of superior dimensions: Thus, suppose the lower one to be a twelve ounce, then the upper one should be a three ounce rocket; the larger one must have a small head formed round its own diameter, and the mouth of the smaller one placed in it; the mouth should be rubbed with mealed powder wetted with spirits of wine; the bore in the charge should not be filled, but have inserted in it a bit of quickmatch, the other end of which should enter the perforations at the top of the greater rocket, which will form a communication between them. The large rocket must be filled only half a diameter above the spindle; if filled higher, it will begin to descend before the upper has taken fire, and produce no additional effect.

The force by which the small rocket goes off, will be sufficient to disengage it from the other, without the use of powder to effect it. One round of paper pasted round the juncture of the two rockets will be sufficient to connect them together. In regard to the rods for towering rockets, the same principles are to be applied as for the others.

3. *Honorary Rockets.*

Take about a pound rocket of our first description, such as is represented in Fig. 18, p. 101; then on the case, close to the top of the rod, tie on, in a transverse direction, a two-ounce case, which should be filled with a strong charge, and choked quite close at both ends; then towards each end, and in the reverse sides, bore a hole of moderate size, and from each carry a leader into the top of the large rocket. When the rocket attains its greatest height, it communicates fire to the cross one at top; from the holes being made in a transverse direction, it will turn round very fast, and represent in its return to the ground, a spiral of descending fire. There are several other methods of adjusting the small case; one consists in letting the rod rise about an inch, or somewhat more, above the top of the rocket, and tying the case to it, so as to rest on the rocket; when adjusted after this manner, the rockets should be without their conical cap.

4. *Caduceus* Rockets.*

If two rockets be fixed obliquely on the opposite sides

* So called from their resemblance (when in action) to the rod borne by Mercury; which, according to fabulous history, was entwisted by two serpents, as the sign and quality of his office, which was given him for his seven-stringed harp.—The term (or Caduce) was also used among the Romans, and applied to the staff or wand of a similar form, carried by those officers who went to proclaim peace with any people with whom they had been at variance.

of a rod, they will form in their flight two spiral lines; they must exactly balance each other on the opposite side of the rod, or they will not rise in a vertical direction. Both ends of the rockets must be choked close, without either head or bounce, for a weight attached to them would obstruct their ascent.

Fig. 23.

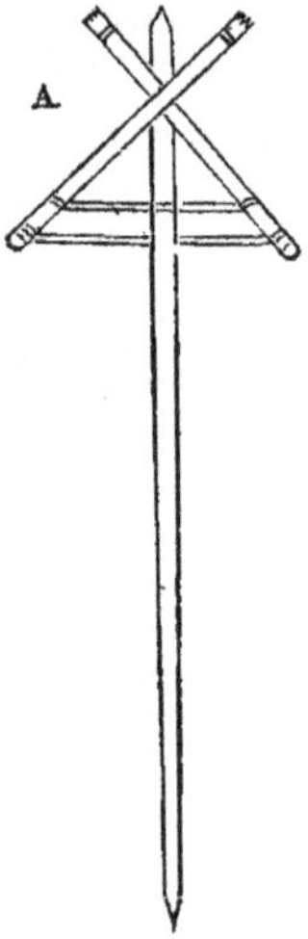

The rod proper for these rockets should be square, and at the top equal to the breadth of a rod for a common single rocket, of the same weight as those intended to be used, and long enough to be in equilibrium, when suspended one length of the rocket from the cross-piece A, Fig. 23, whose length should be equal to about seven diameters of the rocket, and placed about six diameters from the top of the large rod, so that when fixed they will form, with the perpendiculars, an angle of about fifty-five or sixty degrees.

The heads of the rockets should be placed on the opposite sides of the cross-piece, and their ends on the same of the large rod; then their mouths must be connected by a leader, which, when they are fired, must be burned through the middle, and then they will exert their ascending forces at the same time.

5. *Signal Rockets.*

These are of two kinds, namely, those which have reports and those which have not. The first kind may be made somewhat longer in proportion than ordinary, by about one or two diameters, and on their charge must be driven a greater quantity of clay than usual; afterwards their bounce, choke, and cap may be effected after the manner before described.

When of the second kind, their cases and rods must be made very light; in other respects they are similar to the common sky-rocket when without any appendages.

Both the first and latter kind are frequently fired in groups of six, eight, ten, &c.; and considered as signals for the exhibition of pieces of greater magnitude.

When several of these are properly fixed to one rod, and fired together, they form in their flight a most beautiful appearance; for, being so connected, their emissions will unite, and form a tail of stupendous magnitude, and the bursting of so many heads at once, will produce a grand explosion, not unlike (though less productive of injury) the bursting of a balloon in the atmosphere. When rockets are arranged in this manner, particular care must be observed in their filling and ramming, as well as in their exact uniformity of weight, else success is precarious. The rod also must be of proper dimensions, the length of the rods (according to the table) for eight-ounce rockets, which is the best size for this purpose, is six feet six inches; then, if four or six of these be fixed on one rod, the length of it must be about ten feet; in its circumference at top must be made as many grooves as there are to be rockets, and of length to correspond. The rod must be sufficiently large at top to admit the rockets lying close in the grooves without pressing each other too tight.

The rockets must be firmly attached to the rod, or they will be liable by their ascending force to disengage themselves from it; but to prevent this, the best method of fixing them is to let the rod run about two inches above the rockets, which will be sufficient to form a shoulder or stop to each rocket, the groove being discontinued such a distance from the end; when this is done, a little binding round the whole will make it all quite fast. The upper part of the rod may be rounded off in the form of a cone, or, which will be much better, a cap may be pasted over the whole, which (from their meeting with less resistance) will cause them to ascend to a greater height. The rocket being properly fixed, a quickmatch is to be carried from one mouth to the other, which, being burnt in the middle, will communicate immediately to the whole.

When fired, they must be suspended through the rings, as taught in the early part of this article.

6. *Table Rockets.*

This is a simple application of rockets to the spokes of a wheel; to which when fixed they form the felloe.

Their effect (when fired) in the ordinary manner is merely that of revolving on a fixed centre, till their composition is consumed; and by their revolutions representing a vertical or an horizontal circle of fire.

The spokes must be firmly fixed into a block of wood, of lengths, and at a distance from each other, suitable to the lengths of the cases employed. The cases should be those of twelve or sixteen ounces; and filled with the composition given at Nos. 8 or 9, p. 128: they must be carefully rammed.

When the ends of the rockets are fixed to the spokes, which should be notched out proper to receive them and for the purpose of making them more secure, then in the side of each case (outward from the wheel), bore a hole of common dimensions near the clay; these holes should be made in an oblique direction towards the charge, and in and from each must be carried a piece of quickmatch to the centre of the wheel, where they must be tied together and lighted. At the centre of the wheel may be fixed a similar or larger case, which being lighted at the same time, will add much to the exhibition.

The centre of the wheel is frequently fitted to a block of wood, and fired upon a table, when it forms an horizontal wheel; otherwise it revolves on an axis fixed to a post, and in this case a vertical wheel is represented; the centre case may be applied to either.

7. *Scrolls for Rockets.*

These form a pleasing appendage to the heads of rockets that are of considerable magnitude.

They are made in cases about four inches in length,

and their interior diameter about three-eighths of an inch; both ends must be pinched quite close, one before and the other after they are filled; then in the reverse sides make a small vent-hole to the composition, and prime them with mealed powder wetted with spirits of wine.

The heads of rockets may be partly or wholly filled with these cases; when fired they burst quick from their confinement and form a beautiful spiral descent.

The composition may be that of serpents, or the brilliant fire; when either are used it should be prepared strong.

8. *Courantine,* or Line Rockets.*

Among the various modes of exhibiting rockets, none are more pleasing than the present.

Rockets proper for this purpose are those of about half, and three-quarter pounds; they are made after the manner of sky-rockets of the common kind. Any number from one to eight or ten may be used; but five or six will be found to answer best. According to the number of cases used, the courantines are said to be of so many changes. When one, two, or three are only used, they may conveniently be fixed to a small empty case (of the same length), made on a wire former, a little larger than the line on which it is to run, and of considerable substance; but when more than this number are to be used, or a greater change is to be produced, a small perforated cylinder must be procured, of dimensions suitable to the purpose; this cylinder should be of

* From the French term, *courant*, signifying running.

some light wood, such as fine deal, or willow; the perforations must be made exactly through the centre lengthwise. In the same direction, on its circumference, are to be made as many grooves as there are rockets to be employed; in which they must be well secured by tying the whole with a string.

The diameter of this cylinder should be such that, when laid into the groove, the cases may nearly touch each other.

The rockets being all prepared (and their apertures or mouths besprinkled a little with mealed powder and spirits of wine), they are to be laid into the groove, and in such a manner that the head or mouth of the second lies at the same end of the cylinder as the tail of the first; the head of the third the same as the tail of the second; and so on with all the others: they must all be bound tight round with string.

Being thus fixed to the cylinder, from the tail of the first rocket carry a leader to the mouth of the second; from the tail of the second, to the mouth of the third; and so with the whole number, taking care to fix every leader quite secure; and at the same time that the quickmatch does not enter but a very little way into the bore of the rockets, or it will be liable to fire the charge or composition, and thereby destroy all the arrangements.

The runner being now ready for action, a line is to be fixed in an horizontal direction between two posts, or other convenient objects, whose distance from each other (for half-pound rockets) should be about one hundred yards long; this line should be of some strong twine, or (which will answer much better) small brass or iron wire, stretched quite tight between its supports; remembering to put on the runner before fastening both ends. Then (the mouth being next the end of the line) fire the first rocket, which by its force will carry the whole to the end of the line, or nearly. It will be best to have the line too long rather than too short, for if the latter is the case, it will of course make a stand at its extremity, till the remainder of the charge is consumed, which does not look well. But if on the contrary the line is a little too long there will be no such stoppage, not even during the communication of the fire to the next rocket, for the force acquired in its first flight will be sufficient to continue it till such communication is effected; after which it will return in the same manner to the other extremity, and back again in the same order, and so on to the end of the charges arranged on the cylinder.

It is a pleasing exhibition of this kind of fireworks, to arrange them in such manner that when arrived at the extremity of the line, they may communicate fire to some other piece, properly arranged at the end of the line, which in this case should not be so long as before, that the runner may rest a moment before it returns, the better to insure the communication.

To render the runners more agreeable, they are made (of light wood or tin) in the form of different animals, such as serpents, dragons, Mercuries, ships, &c. When thus arranged they are very entertaining, especially when filled with various compositions, such as golden rain, fires of different colors, serpents, port-fires, &c.

The dragons may be made to discharge serpents from their mouths, and two of them arranged on one line, so as to meet each other in the middle, and there appear to contend, till the second case takes fire, when they will run back to the extremity of the line, and then return again with great violence, and produce much amusement both to the operator and spectator.

In the same manner two ships may be represented to contend, and (by filling them well with serpents) be made to pour their broad-sides at each other: or, if they are placed on two separate lines, at a small distance from each other, they may be caused to pass each other in opposite directions; in both cases they will produce a very pleasing appearance.

When the represented animals are made to meet in the middle, the line should be of much greater length, or they will rush together with too great a force.

9. *Revolving Courantines.*

These, while they fly along the line in a straight direction, are, by a simple application of another rocket, caused to revolve, or turn round at the same time. This rotatory motion is easily effected, by fixing to the cases another rocket, which must be placed in a transversal direction; the aperture of which, instead of being at the bottom, like those on the cylinder, must be made in the side, near one of the ends. This transversal rocket must be filled with a very slow charge, or it will be consumed long before those upon the cylinder; when several changes in the runners are intended, two should be fixed in the transversal direction; their diameters should be small, in proportion to their lengths.

The courantines may be made to revolve by other means equally simple and effective. Prepare and fill a case the same as those for Catherine wheels, and wind and tie it nicely round the courantine; this, when lighted with the first case, will cause it to revolve in a very pleasing manner.

When the courantines do not revolve, they may be made to carry on the upper side a jet of fire, or any other ornament which the operator may devise; taking care to suspend, by means of wire, a small weight to the under side, which will keep it always in an erect position.

10. *To represent by Rockets various Forms in the Air.*

To the large case, or head, of about a two pound rocket, place round several small ones of about two or three ounces, the rods of which must be made quite fast to the head, and parallel to the rod of the larger; then, if these be set on fire while the large one is ascending, they will represent in a very pleasing manner, *a tree*, the trunk of which will be the large rocket, and the smaller ones the branches.

If, by means of leaders, the small rockets are caused to take fire when the large one is about half burned in the air, they will represent the form of a comet; and when the large one begins to descend in an inverted position, the small ones will represent a kind of fiery fountain.

If the barrels of some small tubes, or quills, filled with the composition of flying rockets, be placed on a large one, they will, when fire is communicated to them, represent a beautiful shower of fire.

If a number of small serpents be attached to the rocket with a piece of packthread, by the ends that do not catch fire, and if the packthread be suffered to hang down two or three inches between every two, this arrangement, when properly managed, will produce a variety of agreeable and amusing figures.

11. *To cause a Rocket to form an Arc in rising.*

Cut some circles, about three or four inches diameter, out of some tin, or other thin plate; then to the rod of each rocket, and about twice the length of the case from its mouth, fix one of these pieces of tin, nearly at right angles from the rod, and make it quite fast by a bracket underneath. The fire acting upon this, as it proceeds from the mouth of the rocket, will divide the tail in such a manner that it will cause it to proceed in a circular course, and form a very pleasing appearance.

12. *To fire Rockets without Rods.*

Rockets may be made to rise in the air without rods, but in the place of rods they must have attached to them four triangular pasteboard wings, fixed lengthwise on the external of the case, similar to those attached to arrows or darts. The length of these wings should be about three-fourths the length of the rocket; their breadth at bottom should be half their length, and tapered off to nothing at the top. The rocket may be set over a hole in a board, and fired from the under side; or the four wings may rest on four iron pins, six or eight inches in length, drove into a board at suitable distances from each other, and the rocket fired from between them.

Though the greatest care be emplbyed in the exhibition of rockets after this manner, still, their ascent is by far less certain than when a rod is used; therefore the tyro must not be disappointed if he chance to fail of success.

Theory of the Flight of Rockets.

A rocket, being properly constructed, with its rod and other appendages attached, fixed in a vertical position, and fire being applied to its vent, it will (as experience proves) ascend in the air with a prodigious velocity; but upon inquiry into the cause of this ascent, difficulties are met with little contemplated when viewing the beautiful path it described in the medium of its flight.

That this ascent is dependent on the medium (or air) in which it is generated, admits not of a doubt; but to describe how, or in what manner it is effected, has engaged the attention of some of the most eminent philosophers. In consequence, several theories have been advanced for the explication of the phenomena, and among them those of Mariotte and Desaguliers have claimed the most particular attention.

Mariotte attributes the rise of rockets to the resistance, or reaction of the air against the gas, which is generated by the combustion of the composition.

This hypothesis seems to explain the phenomena; but great objections have been brought against it, on account of the difficulty which attends the reduction of it to mathematical investigations. This difficulty arises from the law, which the propelling force must necessarily observe; that is, it will decrease as the velocity increases, in consequence of the partial vacuum left behind the rocket in its flight; so that the velocity becomes as it were both a *datum* and *quæsitum;* and the correct solution of the problem necessarily involves the integration of partial differences of the highest order.

The hypothesis of Desaguliers is somewhat different to the foregoing; it is much more familiar with mathematical investigations, as it reduces the whole theory to the most simple form; and it is not far from being consonant with the known principles of the phenomena, notwithstanding the argument brought against it by Dr. Rees and his editors.

Dr. Desaguliers illustrates his hypothesis in the following manner. Conceive the rocket to have no vent at the choke, and to be set on fire in the conical bore; the consequence would be, either that the rocket would burst in the weakest place, or that, if all parts were equally strong, and able to sustain the impulse of the flame, the rocket would burn out immovable. Now, as the force of the flame is equable, suppose its action downwards, or that upwards, sufficient to lift forty pounds; as these forces are equal, but their directions contrary, they will destroy each other's action.

Imagine then the rocket opened at the vent; by this means the action of the flame downwards is taken away, and there remains a force equal to forty pounds acting upwards, to carry up the rocket, and the stick or rod it is tied to.

Accordingly, if the composition of the rocket be very weak, so as not to give an impulse greater than the weight of the rocket and stick, it does not rise at all; or if the composition be slow, so that a small part of it only kindles at first, the rocket will not rise.

The late Doctor Hutton's philosophy, on the ascent of rockets, is, that at the moment when the powder begins to inflame, its expansion produces a torrent of elastic fluid, which acts in every direction; that is, against the air which escapes from the case, and against the upper part of the rocket; but the resistance of the air is more considerable than the weight of the rocket, on account of the extreme rapidity with which the elastic fluid issues through the neck of the rocket to throw itself downwards, and therefore the rocket ascends by the excess of one of these forces over the other.

This, however, would not be the case, unless the rocket was pierced to a certain depth. A sufficient quantity of elastic fluid would not be produced; for the composition would inflame only in circular coats, of a diameter equal to that of the rocket; and experience shows that this is not sufficient. Recourse then is had to the very ingenious idea of piercing the rocket in a conical hole, which makes the composition burn in conical strata, which have much greater surface, and produce a much greater quantity of inflamed matter and fluid. This expedient was certainly not the work of a moment.

The stick serves to keep it perpendicular; for if the rocket should begin to tumble, moving round a point in the vent, as being the common centre of gravity of rocket and stick, there would be so much friction against the air by the stick, between the centre and the point, and the point would beat against the air with so much velocity, that the reaction of the medium would restore it to its perpendicularity. When the composition is burnt out, and the impulse upwards has ceased, the common centre of gravity is brought lower towards the middle of the stick; by which means the velocity of the point of the stick is decreased, and that at the point of the rocket is increased; so that the whole will fall down, with the rocket end foremost.

During the time the rocket burns, the common centre of gravity is shifting and getting downwards, and still faster and lower as the stick is lighter; so that it sometimes begins to tumble before it is quite burnt out: but when the stick is too heavy, the common centre of gravity will not get so low, but that of the rocket will rise straight, though not so fast.

From the experiments of Mr. Robins, and other gentlemen, it was found that the rockets of two, three, or four inches diameter, rise the highest; and they are found to rise to all heights in the air, from 400 to 1254 yards, which is about three-quarters of mile. For further particulars respecting the theory of the flight of rockets, see Robins's *Tracts*, vol. ii.; *Philosophical Transactions*, vol. xlvi, page 578; and more particularly Mr. W. Moor's *Treatise on the Motion of Rockets*, in which the subject will be found very elegantly treated.

SECTION VII.

TABLES OF VARIOUS COMPOSITIONS.

1. *Serpents.*

Mealed powder one pound, saltpetre one ounce and three quarters, charcoal one ounce.

2. *Pin-Wheels.*

Mealed powder twelve ounces, saltpetre three ounces, sulphur one ounce and a half, steel-filings two ounces.

3. *Common Stars.*

Saltpetre one pound, sulphur four and a half ounces, antimony four ounces, isinglass half an ounce, camphor half an ounce, spirits of wine three quarters of an ounce.

4. *White Stars.*

Mealed powder four ounces, saltpetre twelve ounces, sulphur six ounces and a half, oil of spike two ounces, camphor five ounces.

5. *Blue Stars.*

Mealed powder eight ounces, saltpetre four ounces,

sulphur two and a half ounces, isinglass two ounces, spirits of wine two ounces.

6. *Trailed Stars.*

Mealed powder three ounces, saltpetre one ounce, sulphur three ounces, charcoal one ounce.

7. *Drove Stars.*

Saltpetre one pound, sulphur eight ounces, antimony four ounces.

8. *Pointed Stars.*

Saltpetre eight and a half ounces, sulphur two ounces, antimony one ounce and three-quarters.

9. *Stars of a Fine Color.*

Mealed powder one ounce, saltpetre one ounce, sulphur one ounce, oil of turpentine four drachms, camphor four drachms.

10. *Variegated Stars.*

Mealed powder eight drachms, roch-petre four ounces, vivum two ounces, camphor two ounces.

11. *Brilliant Stars.*

Mealed powder three quarters of an ounce, saltpetre three ounces and a half, sulphur one ounce and a half, spirits of wine one ounce and a quarter.

12. *Trailed Stars.*

Saltpetre four ounces, sulphur six ounces, antimony two ounces, rosin four ounces.

13. *Trailed Stars, with Sparks.*

Mealed powder one ounce, saltpetre one ounce, camphor two ounces.

14. *Gerbes.*

1. Mealed powder one pound and a half, coarse iron-sand five ounces.
2. Mealed powder two pounds, coarse iron-sand eight ounces, saltpetre one pound.

15. *Roman Candles.*

Mealed powder half a pound, saltpetre two pounds and a half, sulphur half a pound, glass-dust half a pound.

16. *Tourbillons.*

For four-ounce cases.—Mealed powder one pound two ounces, charcoal two ounces and a quarter.

For eight-ounce cases.—Mealed powder two pounds, charcoal four ounces and three quarters.

17. *Showers of Fire.*

Chinese.—Mealed powder one pound, sulphur two ounces, iron-sand first order five ounces.

Ancient.—Mealed powder one pound, charcoal two ounces.

Brilliant.—Mealed powder one pound, iron-sand first order four ounces.

18. *Golden Rain.*

1. Mealed powder four ounces, saltpetre one pound, sulphur four ounces, brass-dust one ounce, sawdust two-ounces and a quarter, glass-dust six drachms.

2. Mealed powder twelves ounce, saltpetre two ounces, charcoal four ounces.

3. Saltpetre eight ounces, sulphur two ounces, brass-dust a quarter of an ounce, antimony three quarters of an ounce, sawdust twelve drachms, glass-dust one ounce.

19. *Silver Rain.*

1. Mealed powder two ounces, salpetre four ounces, sulphur two ounces, antimony two ounces, sal-prunella half an ounce.

2. Saltpetre half an ounce, sulphur two ounces, charcoal four ounces.

3. Mealed powder two ounces, saltpetre four ounces, sulphur one ounce, steel-dust three quarters of an ounce.

20. *Water Rockets.*

1. Mealed powder three pounds, saltpetre two pounds, sulphur one pound and a half, charcoal two pounds and a half.

2. Saltpetre one pound, sulphur four pounds and a half, charcoal six pounds.

3. Saltpetre one pound, sulphur four ounces, charcoal twelve ounces.

4. Mealed powder four ounces, saltpetre one pound, sulphur eight ounces and a half, charcoal two ounces.

21. *Sinking Charge for Water Rockets.*

Mealed powder ten ounces, charcoal one ounce.

22. *Water Serpents.*

1. Mealed powder one pound, charcoal one pound.
2. Mealed powder one pound, charcoal nine ounces.

23. *Water Balloons.*

1. Mealed powder two pounds, saltpetre four pounds, sulphur two pounds, antimony four ounces, sawdust four ounces, glass-dust one ounce and a quarter.

2. Mealed powder three pounds, saltpetre four pounds and a half, sulphur one pound and a half, antimony four ounces.

24. *Wheel Cases.*

1. Mealed powder two pounds, saltpetre four ounces, steel-filings six ounces.

2. Mealed powder two pounds, saltpetre twelve ounces, steel-filings three ounces.

3. Mealed powder four pounds, saltpetre one pound, sulphur eight ounces, charcoal four ounces and a half.

4. Mealed powder eight ounces, saltpetre four ounces, sawdust one ounce and a half, charcoal one ounce.

5. Mealed powder twelve ounces, sawdust half an ounce, charcoal one ounce.

6. Saltpetre one pound nine ounces, sulphur four ounces, charcoal four ounces and a half.

25. *Slow Fire for Wheels.*

1. Mealed powder one ounce and a half, sulphur two ounces, saltpetre four ounces.

2. Antimony one ounce six drachms, sulphur one ounce, saltpetre four ounces.

26. *A Dead Fire for Wheels.*

Saltpetre one ounce and a half, sulphur a quarter of an ounce, antimony two drachms, lapis calaminaris a quarter of an ounce.

27. *For Standing or Fixed Cases.*

1. Mealed powder two pounds, saltpetre one pound, sulphur half a pound, charcoal half a pound.

2. Mealed powder one pound, saltpetre half a pound, steel-dust four ounces.

3. Mealed powder ten ounces, charcoal two ounces.

4. Mealed powder half a pound, sulphur two ounces.

5. Mealed powder one pound and a half, sawdust three quarters of an ounce, charcoal two ounces and a half.

28. *For Sun Cases.*

1. Mealed powder two pounds two ounces, saltpetre five ounces, sulphur one ounce, steel-dust twelve ounces.

2. Mealed powder one pound and a half, saltpetre three ounces, steel-dust three ounces and three quarters.

29. *For Spiral Wheels.*

Mealed powder fourteen ounces, saltpetre one pound and a half, sulphur six ounces, glass-dust fourteen ounces.

30. *Globes.*

Saltpetre six ounces, sulphur two pounds, camphor two ounces, antimony four ounces.

31. *Serpents for Pots des Brins.*

Mealed powder ten ounces, saltpetre six ounces, charcoal one ounce and a half.

32. *Fires of different Colors.*

White Fire.—Gunpowder two parts, steel-filings one part : for a pale white add a little camphor. Raspings of ivory give a flame of a silver color, somewhat dazzling to the eyes.

Red Fire.—Gunpowder two parts, iron-sand of the first order one part. Greek pitch produces a flame somewhat red, but more inclined to a bronze color.

Common black pitch produces a dusky flame like a thick smoke, very essential in producing a medium of intolerable obscurity.

Sulphur, mixed in a moderate quantity, makes the flame appear of a blue cast.

Sal-ammoniac and verdigris produce a flame inclined to green.

Raspings of yellow amber give to the flame a lemon color.

Crude antimony a kind of russet color.

33. *For Jets of Fire.*

When the inner diameter of the cases is not more than six lines the following must be the proportions :—

Chinese Fire.—Mealed powder one pound, saltpetre one pound, sulphur eight ounces, charcoal two ounces.

White Fire.—Iron-sand of the first order eight ounces, mealed powder eight ounces, saltpetre one pound, sulphur three ounces, charcoal three ounces.

But when their caliber is from eight to twelve lines, the following are the proportions :—

White Fire.—Mealed powder one pound, saltpetre one pound, sulphur eight ounces, charcoal two ounces.

Chinese Fire.—Saltpetre one pound four ounces, sulphur five ounces, charcoal five ounces, iron-sand of the third order twelve ounces.

Brilliant Fire.—Mealed powder one pound, iron-sand five ounces.

For Jets of larger Dimensions.

Chinese Fire.—Saltpetre one pound four ounces, sulphur seven ounces, charcoal five ounces, and twelve ounces of a compound of the six different kinds of sand.

34. *Sparkling Composition for choked Cases.*

For Black.—Mealed powder and charcoal.

For White.—Saltpetre, sulphur, and orpiment.

For Gray.—Mealed powder, saltpetre, sulphur, and orpiment.

For Red.—Mealed powder, charcoal, and sawdust.

These may be used in any proportion the practitioner may think proper, for a little experience will prove to him that various colors of fire may be produced by only varying the proportions or order of the ingredients, or by rendering them alternately predominant. The same observation will apply to many other cases of a similar nature.

The stars for rockets are prepared by forming the composition into a stiff doughy state, and cutting it into squares like dies, after which they are dusted (while dry) with mealed powder to insure their ignition.

The following are the most approved receipts for the different colors :—

Purple Stars.

	Parts.
Chlorate of potash	42
Saltpetre	22
Sulphur	22½
Black oxide of copper	10
Ethiops mineral	2½

Lilac.

	Pans. Parts.	Stars. Parts.
Potash	49	50
Sulphur	25	25
Chalk	20	22
Black oxide of copper	6½	3

Green.

	Pans or Stars. Parts.
Nitrate of barytes	62½
Sulphur	10½
Potash	23½
Orpiment	1½
Charcoal	1½

Yellow.

Nitrate of soda	74½
Sulphur	19½
Charcoal	6

Crimson.

	Pans. Parts.	Stars. Parts.
Chlorate of potash	4	17
Strontian	67½	55
Charcoal	5½	4
Sulphur	23	18

Great care is necessary in the preparation of the ingredients for colored fires, that they are perfectly dried, otherwise the effect is lost; a sand-bath being the best mode of preparing them. In all cases it is prudent to mix only as much at a time as is necessary, and that as shortly before use as convenient, spontaneous combustion being likely to ensue; and, also, when mixed to be kept in a close-stoppered bottle.

SECTION VIII.

COMPOUND FIREWORKS.

COMPOUND fireworks are those resulting from the combination of the single or more simple kind; principally those which have been already described. The number and variety of figures, and the modification of which they are susceptible, is almost endless, and to describe all, or the greater part of them, would far exceed the limits of this work. It is therefore considered sufficient to select such specimens of simple arrangement as will form a proper introduction to those which are more complex; in which latter case the young pyrotechnist must be left to his own ingenuity, which will readily dictate to him a greater variety than it would be possible to describe.

1. *Girandole* Chests of Serpents.*

The first combination which would naturally suggest itself to the uninformed, is that of a number of serpents so arranged as to take fire all at the same time, and in the end to burst and make a loud report.

---

* A French term, signifying *a cluster.*

This combination is a nest of serpents. The case or box containing them must be made of strong pasteboard, in dimensions equal to the number to be inserted. The piece which forms the top must be perforated in as many places, answering to the number of serpents intended to be fired; they need not be far from each other. At the bottom of the box must be put a little mealed powder for the mouths of the serpents to rest upon, which latter must be rubbed with a little wet mealed powder, in order that they may take fire immediately. To communicate fire to the powder at the bottom of the box, one of the serpent cases must be filled with a slow composition, left open at the top, and inserted about the middle of the box. This case being lighted, it will burn for a short time, or till it reaches the bottom, when a sudden noise will be heard, and all the serpents will be thrown in various directions into the air.

This mode of firing serpents, though puerile in its contrivance, and simple in its production, generally affords much amusement to the spectators; which proceeds principally from the variety of directions given to the serpents; a consequence of their being placed somewhat carelessly in the box, and being ejected at different angles from the same plane.

2. *Girandole Chests of Rockets.*

These chests should be made of some thin boards, in dimensions proportionate to the number of rockets. The rockets best adapted are those of from two to six ounces. The depth of the box should be somewhat more than the length of the rockets with their sticks. The top (being perforated properly to receive the sticks) must be fixed at right angles in the chest, and as far from the top of it as the length of the rocket cases, including the cap, if such are used. The distance between each rocket must be such that they may stand without touching each other. From one hole to another must be cut a groove, sufficiently deep to receive a piece of quick-match, which must be laid from hole to hole in like manner. Below the top, at about two-thirds the length of the rods, must be fixed the bottom, perforated in the same manner, save in the size of the holes, which will be somewhat less owing to the dimensions of the rods. The match being laid as above, take some sky-rockets, and having put a piece of the same match up the cavity of each, left extending a little below the mouth of the rocket, which latter should be rubbed a little with mealed powder, wetted with some liquid before given. The rockets and chest being thus ready, put the rod through the holes in the top and bottom of the chests, in such manner that their mouths may just rest on the quick-match in the grooves, by which all the rockets will be fired at the same time; by lighting any part of the match, it will communicate to the whole of them in an instant. For convenience in placing the rod through the lower holes, a small door should be made in one side of the chest; without this it will be difficult to get the rods in their right places.

Previous to the exhibition of these flights of rockets, they should be covered over, or set in some safe place, or they will be in danger of being set on fire by sparks from other works.

3. *Pots des Brins.*

These are large paper cylinders, filled with powder, stars, sparks, &c. They are generally made of paste-board, and about four diameters long; they should be choked at one end like common cases. They are generally exhibited in numbers, fixed on a plank of some kind, in the following manner: On the under side of the plank, make as many grooves as it is intended to have rows of pots, then at a little distance from each other, and exactly over the grooves, fix as many pegs, about three-fourths or one diameter high; then through the centre of each peg bore a hole down to the groove at bottom, and on every peg fix and glue a pot, the mouth of which must fit tight on the peg; then through all the holes run a quickmatch, one end of which must go into the pot, and the other into the groove, which must have a match laid in it from end to end, and covered with paper, so that when lighted at one end it may discharge the whole almost instantaneously. In each pot put about one ounce of mealed and corn powder; then in some put stars, and in others rain, snakes, serpents, crackers, sparks, &c. When they are loaded, secure their mouths by putting paper over each.

When fired in considerable numbers, these Pots des Brins, from their affording so great a variety of fires, produce a most pleasing exhibition.

4. *Jets of Fire.*

These are a kind of fixed rocket, the effect of which is to throw up into the air jets of fire, similar in some respects to those produced by water. If a number of such rockets be placed horizontally on the same line, it may be easily seen that the fire they emit will nearly resemble a sheet of water, arranging itself in the form of a cascade. When the rockets are arranged in a circular form, like the radii and periphery of a circle, they form what is termed a *fixed sun.*

To procure these jets of fire, the case for brilliant fires must in thickness be equal to a fourth part of the diameter, and for Chinese fires only a sixth part of the same.

The case must be loaded on a nipple, having a point equal in length to the same diameter, and in thickness equal to a fourth part of it; but from the effect of the fire, the mouth generally becomes larger than is requisite; this may be prevented, by charging the case after the manner of the Chinese, who fill it to a height equal to a fourth part of the diameter with clay; this must be rammed down as if it were gunpowder.

When the charge is completed with the composition made choice of, the case must be closed with a tompion of wood, above which it must be choked.

The train or match must be of the same composition as that employed for loading; otherwise the dilatation of the air, contained in the hole made by the spindle, would cause the jet to burst.

Clayed rockets may be pierced with two holes near the neck, in order to have three jets on the same plane. If a kind of top, pierced with a number of holes, be added to them, they will nearly imitate a bubbling fountain.

Jets intended to represent sheets of fire, ought not to be choked. They must be placed in a horizontal position, or inclined a little upwards or downwards.

If to the top of the case be attached a tin cylindrical cap, terminating in a flat, long, narrow mouth (similar to those attached to garden watering-pots), the stream of fire will be very much extended, and the beauty of the exhibition increased. The composition for this article is given in the table, section 7.

5. *Chinese Fountain.*

Provide a piece of dry wood, about six or seven feet long, and about two and a half inches square; at the distance of sixteen inches from the top of this piece (supposing it to be seven feet long, and fixed perpendicular), must be fixed a shelf, sixteen inches long, and in width about two and a half inches, and in thickness about three-quarters. Below this shelf must be fixed three or four other shelves of the same width and thickness, but in length increasing eight inches successively as they go towards the bottom. They must be fixed the same distance from each other as the first one from the top.

Now on the top of the post, insert (into a hole of proper dimensions) a gerbe, or fire-pump; on the first shelf insert after the same manner two gerbes, on the second three, on the third four, on the fourth five, and on the bottom shelf six: they must be so placed, that the next above stand exactly over the middle of the intervals of those below. The gerbes should be placed so that their mouths incline a little forwards; if this be not

Fig 24.

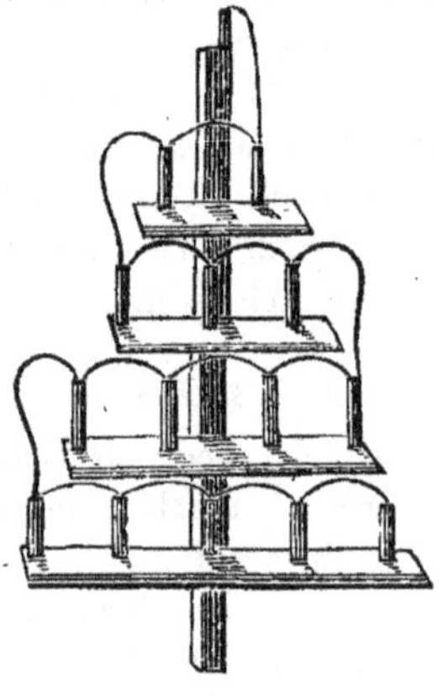

done, the stars thrown out of the cases will strike against the shelf above, and produce but little of that effect, which, when properly arranged, renders them so beautiful.

A proper connection must be formed with your leaders, between the different cases; beginning at the top, and carrying it downward to every one of them. The top one is to be lighted first.

The pyramid, or fountain complete, is represented in Fig. 24.

6. *Pyramid of Flower-Pots.*

In general construction, this article is exactly similar to the one just described; but in place of gerbes or fire-pumps, it is loaded with mortars, filled with serpents, crackers, &c., and having in the centre of each a case filled with spur-fire. The mortars should be made of pasteboard, wound two or three times round a cylinder about four inches diameter, and well secured by glue, by which means their bottoms and tops are fixed to them.

The spur-fire, which is the chief ornament of these pieces, is prepared as follows: It has been said that excellence can never be obtained without overcoming commensurate difficulties; this is certainly verified in the preparation of this composition, for nothing can exceed the difficulty and trouble of preparing it, and nothing can exceed the beauty of its appearance *when properly prepared.* It is said to be the invention of the Chinese, and it is certainly the most beautiful and curious of any yet known.

The principal care in the preparation, is, to have the ingredients of the very best quality; next to that, is the well grinding and mixing them together.

The proportion of the ingredients is saltpetre four pounds and a half, sulphur two pounds, and lampblack one pound eight ounces. One great difficulty is in the mixing these ingredients together; it is best to sift the saltpetre and sulphur together first, and then put them into a marble mortar, and the lampblack with them, which must be worked down by degrees with a wooden pestle, till all the ingredients appear of one color, which will be somewhat grayish, but more inclined to black; when this is done drive a little into a case for trial, and fire it in a dark place; if sparks come out in the form of *stars* or *pinks*, and in clusters, spreading well without any other sparks, it may be considered good: if it appear drossy, and the stars not full, it is not mixed enough; but if the pinks are very small, and soon break, it is indicative of an excess of rubbing; if the excess is great, it will be too fierce, and hardly show any stars: on the other hand, if the rubbing or mixture is defective, it will be too weak, and produce nothing but an obscure or black smoke.

This composition is generally rammed in one or two ounce cases, about five or six inches long; care must be taken not to ram it too hard. The aperture at the choke should not be so wide as is usually given to other choked cases.

It is somewhat remarkable, that the composition should be improved by being kept in the cases; but it is found that they always play better, if suffered to stand a time after they are filled.

In preparing the pyramid of flower-pots, the cases of spur-fire are to be placed in the middle of the mortars, and be connected by leaders, so that they may all be fired together. The cases will first play off in a very pretty manner; and when exhausted, the fire from them communicates to the powder at the bottom of the mortars, and this suddenly taking fire, all blow up simultaneously, and scatter their luminous fragments in the air; the serpents hiss, the crackers bounce, and the illuminated stars fly in all directions, producing considerable amusement and surprise, and forming an excellent conclusion to a small exhibition.

This beautiful composition is also susceptible of other representations, many of which may, without the least danger, be exhibited within a room, as well as in the open air; it is really of so innocent a nature, that it may be (though improperly) called a *cold fire;* for it is found that when well made, the sparks will not burn a handkerchief when held in the midst of them; they may be held in the hand with perfect safety; if the sparks fall a short distance upon the hand, you may feel them like drops of rain.

A pretty exhibition may be produced by placing a number of spur-fires round a transparent pyramid of paper, and fired in a room, or in the open air. In all cases, and every variety of exhibition, this fire is very beautiful, and will always repay the labor of preparation.

7. *Wheels.*

A great variety of forms may be given to this kind of fireworks. They are so called because they are generally made in the shape of wheels, with a nave and spokes radiating from the centre; on the extremities of the latter are adjusted charged cases, of the rocket kind, without heads; in such manner that the tail of one is connected with the head of another, by which method they will take fire successively, and keep up a continued revolution of the apparatus to which they are fixed.

These wheels are either vertical or horizontal, single or double. A single vertical or horizontal may be made after the manner described in Art. 6, Rockets, p. 108.

8. *A Single Horizontal*

May be rendered more pleasing by the following arrangement of the rocket. Provide a wheel, with nave, spindle, and spokes as before; and for the fells a broad cooper's hoop of proper dimensions, nailed on to the end of the spokes, will answer very well. The wheel being thus prepared, the cases must be attached firmly to it, by means of strong packthread, loops passing through the circumference; and in such manner that their heads and tails, as they succeed each other, may alternately incline upwards and downwards, and likewise, when fixed, come very near together.

This being done, from the tail of one case to the mouth of the next following carry a leader, and well secure it by pasting paper round both the joinings: within this pasted paper should be put a little mealed powder, which will serve to blow off the paper, and leave no obstruction to the fire from the cases. To the spindle on which the wheel revolves, fix a case of the same kind as those on the wheel; which must be fired by a leader from the mouth of the last case on the wheel, which case should play downwards. The wheel will be much improved, if instead of a common case in the middle, a case of Chinese fire be fixed, in length sufficient to burn as long as three cases on the wheel. In all the cases (except the first one), on each wheel, should be driven a ladleful or two of slow fire, in any part of the case: in the end of one or two alternate cases ram a ladleful of dead fire composition, which must be very lightly rammed; many other changes in the appearance may be produced by alternately ramming in composition of different orders.

Horizontal wheels are frequently fired two or three

Fig. 25.

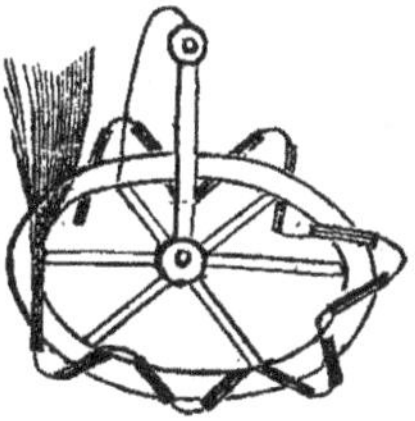

at a time; and being prepared after the same manner, will keep time with each other: when thus arranged, the slow or dead fire is omitted. These wheels may be made from ten to twenty inches diameter.

A horizontal wheel, with the cases fixed, is represented in Fig. 25.

9. *Plural Wheels.*

So called from their being several of them fixed on the same axis; they are generally horizontal, and in number three. The diameter of the middle wheel may be a little less than the other two.

The cases must be fixed to the ends of the spokes in notches cut on purpose, or there may be half cylinders of tin nailed to the ends of the spokes, and the cases tied in them. The bottom cases should play obliquely upwards; the middle set horizontally; and the upper cases obliquely downwards. The leaders must be arranged so that the cases may burn first up, then down, then horizontal, through the whole sets. By driving in the end of the last case two or three ladlesful of slow fire, it will cause it to burn till the wheel has stopped its course; and if the other cases are fixed in the contrary way, the wheel will then revolve in a contrary direction, and have a pleasing appearance. For the case at top of the axis, a gerbe may be well employed; the case on the spokes should be filled with a strong brilliant charge.

10. *Spiral Wheels.*

These in their principal construction differ but little from the foregoing; the following are the principal differences: The nave should be about seven inches long; instead of a spindle at top, make a hole for the case to be fixed in; in the nave must be fixed two sets of spokes near the top and bottom; the spokes should not be more than about three inches long; the cases must be placed in such a manner, that those at top play downwards, and those at bottom play upwards, but the third or fourth case must play horizontally. The case in the middle may begin with any of the others; six spokes will be sufficient for each set, by which the wheel may contain twelve cases, besides the top one: the cases should be about seven inches in length.

11. *Illuminated Spiral Wheels.*

Provide a horizontal wheel with circular fells complete, which should be about two feet six inches diameter; on its circumference, and at equal distances from each other, fix three pieces of light deal about four feet long, and at top connect them to a cylindrical block about three inches diameter; this block must be perpendicular to that of the wheel below. The wheel being thus far advanced, have a thin flexible lath or hoop, and having nailed one end to the bottom of one of the upright pieces, proceed to wind it round the three uprights in a spiral line from the wheel to the top block, to which the other end must be made fast; on the top of the block fix a case of Chinese fire; on the wheel you may place any number of cases, which should incline downwards, and burn two at a time. If the wheel has ten cases, the illuminations and Chinese fire may begin with the second case.

The axis for this wheel must pass through the bottom nave, and into the block at top.

This wheel may be easily wrought into a double spiral wheel, by winding round it another lath in an opposite direction, and clothing it in a similar manner. At the top of either may be placed a case of spur-fire, or amber light, or any other article the pyrotechnist may think proper.

12. *Balloon Wheels.*

These are horizontal wheels, generally made of solid one-inch elm board, about two feet six inches diameter. On the top arrange and fix in pots, three inches diameter and about six inches high, in number equal to the cases on the wheel: near the bottom of each pot make a small vent, into each of which carry a leader from the tail of each case. The pots may be loaded with stars, crackers, serpents, &c. As the wheels turn, the pots will be successively fired, and caused to throw into the air a great variety of fires, which, taking numerous and various directions, will present a pleasing exhibition.

13. *Ground Wheels.*

These are of a very simple contrivance. Provide two light wheels, in diameter from two to three or four feet; they must be fixed firmly to a square axletree, or in such manner that they cannot revolve on it; the axle may be about three feet in length. Then on the middle of this axle, is to be firmly fixed a fire-wheel, which must be so much less in diameter, that when the cases are attached to it, it may be quite clear of the ground; care must be taken that this middle wheel be fixed at right angles from the axle, or it will not keep in a straight direction when set in motion. Now the first case being fired, it is evident that motion will be given to the fire-wheel, and this being fixed firmly to the axle of the others, the consequence that follows will be to give absolute* motion to the whole apparatus; which, if placed on level ground, will proceed to a distance proportionate to the number and strength of the eases employed.

By attaching a second set of cases, so arranged as to take fire when the first set are consumed, the wheel (running upon level ground) will return to the same place from which it received its primary impulse.

This kind of wheels, when constructed with care, affords a very pleasing recreation. It may be readily seen, that many other ornamental pieces of less magnitude may be attached to the same axle: a good level school-ground is favorable for the exhibition of this article.

14. *Horizontal changed to a Vertical Wheel.*

The wheel for this should be about three feet six inches in diameter. On its circumference fix sixteen half-pound cases filled with brilliant charge, two of which should burn at a time. On each end of the nave, must be a tube or barrel of tin or brass, in diameter something less than that of the nave, and in height about six inches; this is the construction of the wheel. The stand to which it is to be fixed is as follows: set a post of any kind of wood, about four inches square, firmly into the ground, standing up about five feet; then from

* *Absolute* motion is the change of absolute space or place of bodies, as the *motion of a projectile*, the flight of a bird, or the motion of our own apparatus.

the top saw off about two feet, which piece must be joined again at the place where it was cut, with a strong hinge on one side, in such a manner that it may lift up and down in front of the stand; on the top of the bottom part, the side on which the movable part falls, fix a very strong bracket projecting about a foot from the post; and at the extremity of this form a tenon, corresponding to a mortise made in the movable part, so that when it falls it may be firmly fixed to it; this particular must be attended to, or the force with which the wheel revolves when vertical, will be liable to pull off the hinge. On the side of the short post opposite the hinge, nail a piece of wood, extending about eighteen inches down the bottom part of the post, to which it must be tied with a piece of string only; this will be sufficient to keep the short part perpendicular; in the top of the latter, fix a spindle, ten or twelve inches long; on this spindle put a wheel; then fix on a brilliant sun with a single glory, the diameter of which must be about six inches less than that of the wheel. The wheel being ready to fire, light the wheel part first, and let it run horizontally till four cases are consumed, then from the end of the fourth case carry a leader into the tin barrel that turns over the end of the stand; this leader must be met by another brought through the top of the post, from a case filled with a strong port-fire charge, and tied to the bottom post, with its mouth directed towards the string which holds up the upper part of the post, so that when this case is lighted, it will burn the string and let the wheel fall downward, by which means it will become vertical; then from the last case of the wheel carry a leader into the barrel next the sun, which will exhibit its beauties immediately the wheel has ceased.

The sudden change of this piece renders it very surprising and pleasing to the observers of it, and entitles it to great attention.

15. *Vertical Scroll Wheel.*

A wheel of this kind may be made of any diameter. The nave may be of moderate size, and to it fix four or six spokes, at right angles to each other, and united to the fell or circumference; round the latter fix any num-

Fig. 26.

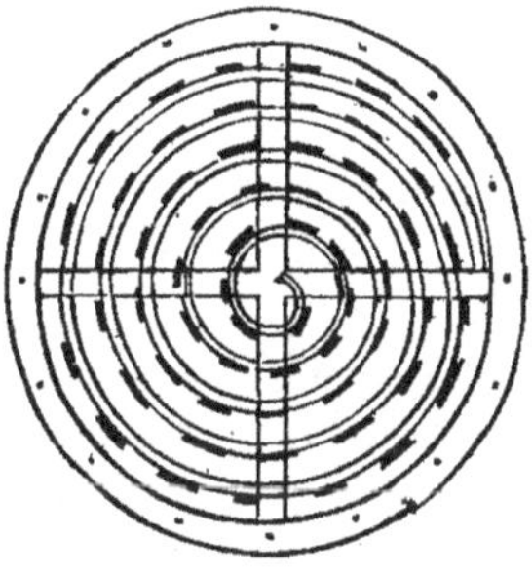

ber of charged cases: on the front of the spokes form, with some strong iron wire, a scroll, or volute, in dimensions proportionate to the wheel, beginning at the centre; on this scroll tie cases of brilliant fire, which should not be too large, and placed head to tail, as in other similar arrangements. The case nearest the circumference must be fired first, which being farthest from the centre has most power to set the wheel in motion. The charged cases may be before, at the same time, or after the scroll.

This wheel may be wrought into one far more ornamental and complex. A double scroll might be formed on the spokes, as well as a double set of charged cases on the circumference; a pot of some kind at the centre would readily suggest itself.

Remarks on Wheels.

In all articles of the wheel kind, the tyro must observe to increase the strength of his composition for cases, as his wheels increase in diameter; for a rocket proper for a twenty-four inch wheel will not do well for one that is much larger.

The following rule as to this particular may serve in many cases: divide the diameter of the wheels, taken in inches, into three parts, and it will give the length of the cases, and generally within one, the number it will require to go round it. Thus suppose the wheel is twenty-four inches diameter, divide by 3; $\frac{}{3}$ equals 8, which is about the length of the cases: and 7 : 22 : : 24 : 528 which divided $\frac{528}{7} = 75.3$ and $\frac{75}{8} = 10$ equal 10, the number of eight-inch cases it will take to go round the circumference.

This is not given as a particular but as a general rule,

or one that will assist a little in the arrangement of these articles.

16. *To represent the Fir-Tree.*

Provide a post six or seven feet long and three inches square; then on the far side, at nine inches from the top, fix in four short pegs to fit the inside of the cases; nine inches from these fix similar pegs; nine inches lower fix others similar to the last; and from these, the same distance, fix other pegs; all these four sets must incline upwards; below them, at the same distance, must be fixed another set inclining downwards; the angles of inclination in all may be about forty-five degrees from the upright post. At the top of the post place a four-inch mortar loaded with stars, rains, crackers, &c. In the middle of this mortar place a case with any sort of charge fired with the others, which should be filled with a brilliant charge. The tree may be made of any size, and the other ornaments made use of as suits the operator.

17. *Yew-Tree of brilliant Fire.*

Provide a piece of wood about four feet long, two inches wide, and one thick; at top, on the flat side, fix a hoop about fourteen inches diameter; and round its edge and front place illuminations, and in the centre a five-pointed star; then on each side, at about eighteen inches from the edge of the hoop, place two twelve-inch cases of brilliant fire; below which place two more cases of the same size, and at such a distance that their mouths may almost meet them at top; then close to the ends of these fix two more of the same cases, which must stand parallel to the others. The cases being thus fixed, the leaders must be applied in such a manner that the illuminations and stars at top may all take fire at the same time. Fig. 27 represents the arrangement of the article.

Fig. 27.

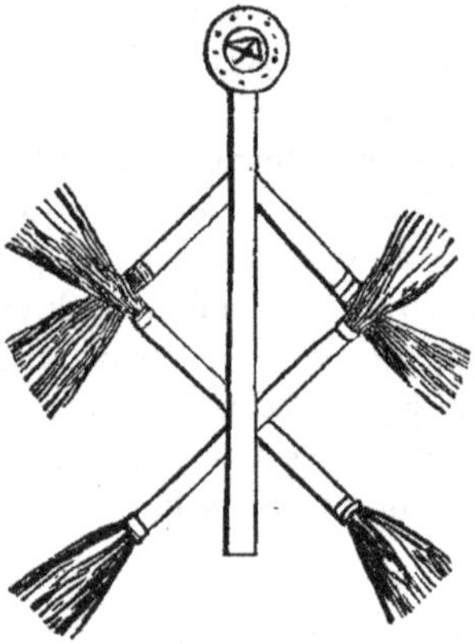

18. *Fixed Fire Globes.*

These articles are divided into two kinds, one with projected cases, the other when the cases are concealed.

For a globe with concealed cases, provide a spherical globe of any diameter; divide the surface in fourteen equal parts, and at each division bore a hole perpendicular to the centre: in every hole except one (which must be reserved for the spindle, on which it must be fixed) insert a case filled with brilliant or any other charge; the mouths of the cases must be even with the surface of the globe; from the mouths of the several cases must be cut a groove, and in it laid a leader, for the purpose of firing them altogether. The globe must be covered over with paper, and painted in what manner the tyro thinks proper. When dry, it is to be fixed upon the spindle, and it is ready for exhibition.

For projected Cases.—The preparation is nearly the same; the difference being only to let each case project from the globe about half or two-thirds of its length; their mouths are to be connected by leaders for the same purpose as before, and exhibited in the same manner.

19. *Globes which leap or roll on the Ground.*

Construct a hollow wooden globe of any dimensions at pleasure; it must be very round, both internal and external; its thickness must be equal to about the ninth part of its diameter. In this globe insert a small wooden cylinder (A, Fig. 28) in breadth equal about one-

Fig. 28.

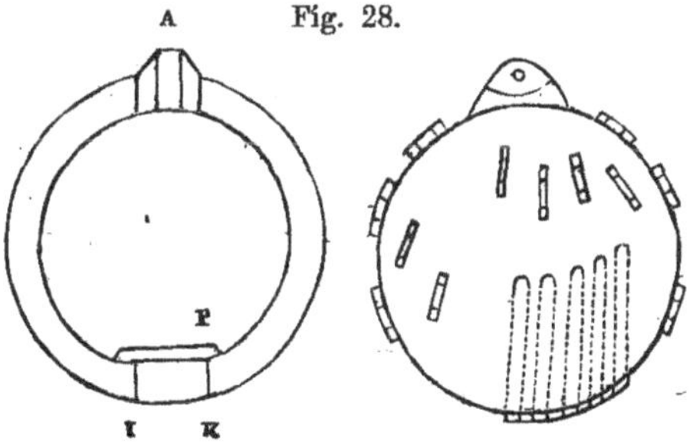

fifth diameter of the globe, its thickness about half that of ditto; of the same size and opposite to this cylinder must be another aperture. It is through this latter aperture that fire is communicated to the globe, when it has been filled with proper composition through the lower end of it; and by which there is the convenience of filling, and of putting, as is generally done, a petard or report of metal, filled with good grain powder over the inside of the aperture; besides this petard four or five others of a similar nature, only they need not be in metallic cases, are to be inserted; they must be loaded with good grain powder filled to their orifices. The composition for filling the remaining cavity of the globe is—one pound of bruised gunpowder, six pounds of saltpetre, three pounds of sulphur, two pounds of iron-filings, and half a pound of Greek pitch. This composition will not require much grinding or sifting; it will be sufficient if the different ingredients be well incorporated. It should not be made up quite dry, but with a little of one of the liquids before mentioned.

A globe prepared as above, on being fired by means of a match attached to the orifice A, will leap and bound about as it burns, or according to the accidental explosion of the petards, which are set on fire by the composition.

Instead of placing these petards in the inside, they may be affixed to the exterior surface of the globe, which they will make to roll and leap about as they successively take fire. They may be arranged in any manner on the surface of the globe, provided a connection is formed between them by means of leaders.

Many differences in the arrangement and form of these globes may be made, and which will readily suggest themselves to the ingenious practitioner; such as an arrangement of rockets on the inside, laid head and tail together; but in case the globe should be of paper or pasteboard, made in two equal hemispheres, and joined together by paper, the match must be applied through a hole in the globe made opposite the mouth of the first rocket; these rockets should have no petards at their tops. It must here be observed that the globes must be perforated in various parts, otherwise they will burst by the combustion of the composition.

When used as water globes, care must be taken to seal and plug up the lower aperture I, K, first with a tompion, or plug of wood, and afterwards with some melted pitch; which latter may be put all over the globe in order to preserve it from the water. Over the plug at bottom, and previous to the application of the pitch, must be melted such a quantity of lead, as will cause the globe to sink in water, till nothing but the part A will remain above its surface; this will be the case when the weight of the globe and its contents, with the lead attached, becomes equal to the weight of an equal volume of water. If the globe be then placed in the water, the lead, by its superior gravity, will make the aperture I, K, tend directly downwards, and keep in a perpendicular position the cylinder A, to which fire must have been previously applied.

Trial should be made respecting the quantity of lead previous to its being exhibited, which may be easily done. The figures referred to represent a globe under various modes of arrangement.

20. *Moon and Seven Stars.*

Provide a circular board about five feet diameter; and out of the middle cut a piece about fourteen inches diameter; then, over the opening, put a piece of white Persian silk, on which paint a moon's face; over the whole of the large board draw a seven-pointed star, terminating in the circumference; then, on the lines forming the star, bore a number of holes at small distances from each other, wherein fix pointed stars. In each of the spaces between the points of this large star cut out a five-pointed star, and cover each with oiled silk.

When this is to be exhibited, fix it on a spindle in front of a post, with a wheel of brilliant fire behind the face of it; so that while the wheel burns, the moon and stars will appear transparent; and when the wheel has ceased burning, they will disappear, and the large star in front, formed of the pointed stars, will begin, being lighted by a pipe of communication from the last case of the vertical wheel behind the moon, which must be effected as taught in a foregoing article.

21. *Suns Fixed and Movable.*

Among the various amusing articles of pyrotechnic produce, none are more beautiful or afford greater remuneration of pleasure, than those under the denomination of suns. They are of several kinds, as fixed, movable, and transparent; they are all of simple construction.

Fixed suns after the following manner: Provide a nave of wood, and in it fix fourteen or sixteen pieces in the form of radii; and to these radii attach jets of fire, the mouths of the jets being towards the circumference. A match must be applied, in such a manner that the fire communicated at the centre may be conveyed, at the same time, to the mouths of each of the jets; by which means each throwing its fire, the appearance will be that of a radiating sun; the wheel must be fixed in a vertical position.

The jets may be so arranged as to cross each other in an angular manner; in which case, instead of a sun you will have a star, or a sort of cross resembling that of Malta. Some of these suns are made also with several rows of jets; when they are so arranged they are called *glories*.

The wheel, or sun, may be caused to revolve by attaching jets to it in the direction of the circumference, with their heads and tails together. When the wheel is heavy, four of the rockets must be fired together, and this in the following manner: Supposing there are twenty cases employed, fire must be communicated, at the same time, to the first, the sixth, the eleventh, and the sixteenth; from which it will proceed to the second, the seventh, the twelfth, the seventeenth, and so on. These four rockets will make the wheel turn round with rapidity.

If two similar suns, with horizontal axes, are placed one behind the other, and made to turn in opposite directions, they will produce a very pleasing effect of cross-fire.

Three or four suns arranged on a similar axis, might be implanted in a vertical one, movable in the middle of a table; which, revolving around it, would seem to pursue each other. They must be fixed firm on their axis and this axis must turn in the upright one in the middle of the table; and the place where they rest on the table should be furnished with a very movable roller.

For a transparent sun, a prepared face of oiled paper, or Persian silk, painted in a proper manner, must be provided, and strained tight upon a hoop, which must be supported by pieces of strong wire, six or seven inches from the wheel, so that the light of it may illuminate the face. After the same manner may be represented in the front of a sun the words *Farewell*, *Vivat Regina*, or *Apollo*, or any other figure may be painted on the silk.

Sometimes a small hexagonal wheel is fixed to the nave of the large wheel, whose cases must be filled with the same charge as the other; two of which must burn at a time, and begin with the others.

For a sun five feet in diameter, the cases should be those of eight ounces, filled with composition about ten inches. If the wheels are larger, the cases must be proportionate to them.

22. *Composition for representing Animals and other Devices in Fire.*

Reduce some sulphur to an impalpable powder, and having with starch formed it into a paste, cover with it the figure you intend to represent on fire; the figure must be first coated over with clay, to prevent it from being burnt.

When the figure has been covered with this paste, sprinkle it a little, while still moist, with pulverized gunpowder; and when the whole is perfectly dry, arrange some small matches on the principal parts of it, that the fire may be speedily communicated to it on all sides.

By the same method may be formed festoons, garlands, and other ornaments, the flowers of which might be imitated by fire of different colors, and arranged on any plastered architecture.

Aquatic Fireworks.

Though fire and water are of very opposite natures, yet there are many fireworks which will burn and produce their effect even when immersed in their opposite element; of these, rockets are the most pleasing. They may be made from four ounces to two pounds. The cases are made as those for sky-rockets, differing only as it regards the thickness of them, which should be somewhat greater, and in the manner of filling, which latter is the most particular, requiring a variety of compositions rammed in alternate layers, for the purpose of making them alternately dive and swim. The compositions are chiefly of three kinds, namely, that of sulphur two ounces, saltpetre four ounces, and mealed powder one ounce and a half, and about one fourth of an ounce of antimony. The second kind, called a *sinking charge*, is composed of eight ounces of mealed powder, and three-fourths of an ounce of charcoal. The third, called *an ordinary charge*, composed of mealed powder, saltpetre, sulphur, and charcoal, varied as in the following proportions: sometimes a small portion of seacoal or sawdust is blended with them.

1. Mealed powder six pounds, saltpetre three pounds, charcoal five pounds.

2. Mealed powder four pounds, saltpetre four pounds, sulphur two pounds.

3. Mealed powder four ounces, saltpetre one pound, sulphur eight ounces and a half, charcoal two ounces.

4. Mealed powder one pound, saltpetre three pounds, sulphur one pound, charcoal nine ounces.

5. Saltpetre one pound, sulphur four ounces and a half, charcoal six ounces.

In filling, one ladleful of slow-fire is first rammed in the case; then one or two of the sinking charge; the common and the sinking charge are placed alternately to within about two diameters of the top. Over the last layer is placed one ladleful of dry clay; and a perforation made through into the charge. The remainder of the case upwards, to within about half a diameter, is filled with corn powder, and two or three folds of the paper turned over it; then the ramming paper at the end is secured with strong thread, and afterwards dipped in melted pitch or wax. When several rockets are thrown into the water at the same time, care should be taken to select those which have been filled and rammed in an uniform manner. To secure the cases from the action of the water, it is obvious that they should be prepared as such; this is done by varnishing them over with linseed oil or common varnish.

Leaders, and pipes of communication, must be prepared after the same manner as rockets, as far as regards their cases; that is, they must be made somewhat stronger, and, when fixed, to be varnished over as before; taking care not to varnish till all your pasting is completed.

To make a Fire Fountain for the Water.

Provide a circular float three feet in diameter; in the middle fix a round post four feet high, about two inches diameter; round this post fix three circular wheels made of thin wood. Place the largest within two or three inches of the bottom, which should be not much less than the float. The second wheel must be about two feet two, and fixed two feet from the first. The third wheel must be sixteen inches diameter, and fixed within six inches of the top of the post. Then take eighteen four or eight ounce cases of brilliant fire, and place them round the first wheel with their mouths upwards, and inclining downwards; on the second wheel place thirteen cases, in the same manner as those on the first; on the third place eight more in the same manner as before; and on the top of the post fix a gerbe; then clothe the cases with leaders, so that both they and the gerbes may take fire at the same time. Before firing this work it is best to try it in the water, to see whether the float is properly made, so as to keep the fountain upright.

Aquatic exhibitions are almost as numerous as those of the other kind, but it is not necessary to describe more, as so many of them depend on the taste and ingenuity of the practitioner.

In closing this work, it may be well to direct the attention of the tyro to the frequent displays of fireworks which may be witnessed at the several places of amusement in and about the metropolis, particularly at the public gardens, as he will there see many curious and beautiful devices, which may serve as specimens for him to exercise his talent and ingenuity upon. The theatres also often present very excellent exhibitions of the pyrotechnic art, which are well worthy of attention, and all tend to show that it is far from being on the decline in England.

Red. Fire

Nit. of Strontia ʒ xvi
Chlor. of Kali ℥ viii
Shelack ℥ iii

Pulverize each article separately and very fine then mix and sift

J.G. Knighton

Red Fire.

Nit. Strontia
Chlor. Kali ℥ xxii
Shellac ℥ xvi
Lycapodium ℥ viii
℥ i

Crimson fire.

Nit. Strontia ℥ Lxxiss.
Chlor. Kali ℥ ivss
Carbon ℥ vss
Sulphur ℥ xxiiss

Green fire.

Nit. Barytae ℥ xvi
Chlor. Kali ℥ viii
Shellac ℥ iv
Lycapodium ʒ iv

Blue fire.

Nit. Kali ℥ viii
Sulphur ℥ iv
Sulph. Nig. Stibium ℥ i

Yellow fire.

Nit. Natrum (dry) ℥ xivss
Sulphur ℥ xixss
Carbon ℥ vi

White fire (without Smell)

Nit: Natrum ℥ xxiv
Chlor. Kali ℥ i
Lycapodium ℥ i
Stibium Nig. Sulph. ʒ iv
Rosin alb. ʒ ii

Shower fire.

Sulphur Dep. ʒ iv
Nit: Kali ʒ ii
Stibium Sulph. Nig. ℥ ii
Carbon ℥ i
Ferri filum ℥ i

N.B. Stone coal makes mor spark

Yellowish-red fire

Natrum Nit: ℥ v
Carbon ℥ i
Sulphur Dep. ℥ i

White Bengal fire

Kali Nit. ℥xii
Sulphur Dep. ℥viii
Stibium Sulph Nig ℥iv

Green fire

Nit. Baryta ℥ii
Kali Oxymuriat ℥i
Sulphur Dep. ℥i

Red Bengal fire

Nit: Strontia ʒxi
Kali Chlor. grs lviii
Carbon ʒi
Pyria Pulv. ʒiss
Sulphur Dep. ʒiii

White fire. Ind.

Nitre. ℥xxiv
Sulphur ℥vi–vii
Realgar ℥ii–iii

Fire

Nitre Dep:	℥ xv
Sulphur Dep.	℥ xv
Oil Lini	℥ x
Gun Powder	℥ xxx
Camphor	℥ ii
Gum Acacia	℥ iv
Alcohol	℥ viii

Desolve the Camphor and Nitre in the Alcohol. Desolve the powder in a little water with the gum. Then mix the sulphur wit the oil; mix the whole thoro -ghly roal it out and dry cut in small peices

Silver Rain

Mealed Powder ℥ x

Sulph Dep. ʒ iv

Kali Nit. Dep. ʒ iv

Steel filings ℥ ii

Chinees Silver Rain

Mealed Powder ℥ xvi

Sulphur Dep. ℥ ii

Kali Nit. ℥ i

Brass filings ℥ iss

Shower fire

Kali Nit. ʒ iv ss

Sulphur Dep. ℥ ii

Lamp Black ℥ iss

White fire

Mealed Powder ʒ x

Steel filings ℥ iv

Greek fire

Nitre Dep.	℥ vi
Sulphur Dep.	℥ ii
Antimon. Sulph.	℥ i
Camphor	ʒ ii
Pyrio Pulv	ʒ iv

Red Fire.

Nit. Strontia	℥ XVI
Chlor. of Potash	℥ III
if Strontia is <u>not</u> good	
if good	℥ I ss.
Shellac	℥ IV

Strontia	℥ X
Shellac	℥ II
Chlor. Potash	℥ I ss.
Lycapodium	ʒ II

Made in the USA
Monee, IL
04 November 2026

67378958R00115